4-H Robotics: Engineering for Today and Tomorrow

Junk Drawer Robotics

Level 2

Robots on the Move

Presenter's Activity Guide

Junk Drawer Robotics Level 2
Robots on the Move

Table of Contents

Introduction to Junk Drawer Robotics 3
What Is 4-H Science? 4
Experiential Learning Process 5
Positive Youth Development 6
4-H Life Skills 7
What You Will Need for Junk Drawer Robotics 8
Focus for Robots on the Move 10

Overview of Module 1: Get Things Rolling 13
- Activity A — Slip N Slide 16
- Activity B — Rolling Along 18
- Activity C — Clipmobile Design Team 20
- Activity D — Clipmobile Build Team 27

Overview of Module 2: Watt's Up? 31
- Activity E — Light Up My Life 34
- Activity F — Magnetic North 36
- Activity G — Can-Can Robot Design Team 38
- Activity H — Can-Can Robot Build Team 40

Overview of Module 3: Get a Move On 43
- Activity I — Gear We Go Again 46
- Activity J — Gears and More Gears 49
- Activity K — Gear Train Design Team 53
- Activity L — Gear Train Build Team 55
- Activity M — Es-Car-Go Design Team 57
- Activity N — Es-Car-Go Build Team 59

Overview of Module 4: Under the Sea ROV 61
- Activity O — Pennies in a Boat 64
- Activity P — Sink or Float 66
- Activity Q — Sea Hunt Design Team 69
- Activity R — Sea Hunt Build Team 71
- Activity S — To Make the Best Better Design and Build Team 73

Glossary 76
References 77

Introduction to Junk Drawer Robotics

The goal of *4-H Junk Drawer Robotics* is to make science, engineering, and technology engaging and meaningful in the lives of young people. The activities do this by encouraging youth to use the processes and approaches of *science*; the planning and conceptual design of *engineering*; and the application of *technology* in their personal portfolios of skills and abilities.

The *Junk Drawer Robotics* curriculum is divided into three levels or books each around a central theme related to robotics design, use, construction, and control. Each level starts out with background information on working with youth, curriculum elements, and a focus on the concepts to be addressed.

- In Book 1, the theme is robotic arms, hands, and grippers.
- In Book 2, the theme is moving, power transfer, and locomotion.
- In Book 3, the theme is the connection between mechanical and electronic elements.

Modules within each book target major concepts of the theme. The modules each contain five or more activities that help the members develop an understanding of the concepts, create solutions to challenges, and develop skills in constructing alternatives.

Your role as a presenter in this curriculum is different from the normal role of the teacher that we may know from a school setting. It is not about mere transfer of knowledge from teacher to student. It is about assisting learners in developing their own knowledge and Problem solving skills. This is done by bringing together a scientific inquiry and engineering design approach to learning. Youth will learn about a topic by exploration. When given a problem, they will design a solution. Then, using what they have learned and designed, they will build or construct a working model.

The presenters of this curriculum may be teachers, after-school staff, or volunteers, including teens working with younger youth. In the case of teens, an adult coach or mentor could provide support, training, and guidance to teams of two or three teens who present the activities together. As a presenter, you will assist the youth in understanding the processes of science, engineering, and technology by how you ask questions and have them share their ideas, designs, and results.

The robotics curriculum is designed around three themes of science, engineering, and technology. Each module has activities in each of these three areas. As a framework, *4-H Junk Drawer Robotics* uses these simple definitions developed by Anne Mahacek, former 4-H teen member who is now a mechanical engineer and grad student in mechatronics.

To Learn: **Science** is finding out how things work.

To Do: **Engineering** is using what you found out to design something to work.

To Make: **Technology** is using tools and processes to make something work.

What Is 4-H Science?

The National 4-H Science initiative addresses America's critical need for more scientists and engineers by engaging youth in activities and projects that combine nonformal education with design challenges and hands-on, inquiry-based learning in a positive youth development setting. These experiences engage youth and help them build knowledge, skills, and abilities in science, engineering, mathematics, and technology.

All 4-H Robotics activities and projects are:

- Based on National Science Education Standards and Standards for Technological Literacy,
- Focused on developing abilities in science, mathematics, engineering, and technology,
- Led using the Experiential Learning Model,
- Tied to developing Life Skills for youth, and
- Delivered in a positive youth development context by trained and caring adults.

National Science Education Standards (NSES)

The National Science Education Standards present a vision of a scientifically literate person and present criteria for our education system that will allow that vision to become reality (National Research Council, 1996). NSES outline what students should know, understand, and be able to do as they progress through their science education. Emphasis has shifted from being solely on "the content to be learned" to include "**how** students learn" and "**how** the content is presented."

Scientific Inquiry refers to the diverse ways in which scientists study the natural world and propose explanations of the world based on the evidence derived from their work (National Research Council, 1996). But scientific inquiry is not limited to the work of scientists. Young people driven by curiosity and given a structure can pose questions, make observations, analyze data, and offer their own explanations. Supporting youth in developing the skills and understanding necessary to engage in scientific inquiry is a central focus of the NSES and 4-H Science.

Standards for Technological Literacy (STL)

Standards for Technological Literacy define what youth should know and be able to do in order to be technologically literate (International Technology Education Association, 2000). Technological literacy is important to all youth whatever path they pursue in life. They offer a common set of expectations for what students should learn in the study of technology and what is developmentally appropriate at different ages. The 20 standards address five areas: the nature of technology, technology in society, design, abilities for a technological world, and understanding the designed world. The activities in *Junk Drawer Robotics* help students develop literacy in each of these areas.

Science, Engineering and Technology Abilities

According to Horton, Gogolski, and Warkentien (2007), effective teaching in science, engineering, and technology must focus on how youth learn the content **and** how the material is taught. Based on a review of the science, engineering, and technology education literature, the authors set forth 30 important processes and refer to them as Science, Engineering and Technology Abilities (Horton, Gogolski, & Warkentien, 2007). This set of science, engineering, and technology life skill outcomes is emphasized throughout the 4-H Robotics curriculum. Examples of some of the 30 abilities that are developed in this curriculum include: observe, categorize, organize, infer, question, predict, evaluate, use tools, measure, test, redesign, collaborate, summarize, and compare. Each Module in *Junk Drawer Robotics* identifies particular abilities in science, mathematics, engineering, and technology that are focused on in that module.

Experiential Learning Process

Experiential learning allows young people to create or develop their own answer to a question instead of repeating "the answers" (Maxa, et. al., 2003). In the experiential learning process, youth are encouraged to think, explore, question, and develop decisions. Two important components of the experiential learning process are a period of reflection, during which the learner shares and processes the experience, and the application of new learning in "real world" situations. The experiential learning model contains five steps: experience (doing), share (what happened?), process (what's important), generalize (so what?), and apply (now what?) (Maxa, et. al., 2003).

The Experiential Model

This model helps leaders formulate activities to reflect the DO, REFLECT, APPLY in five steps.

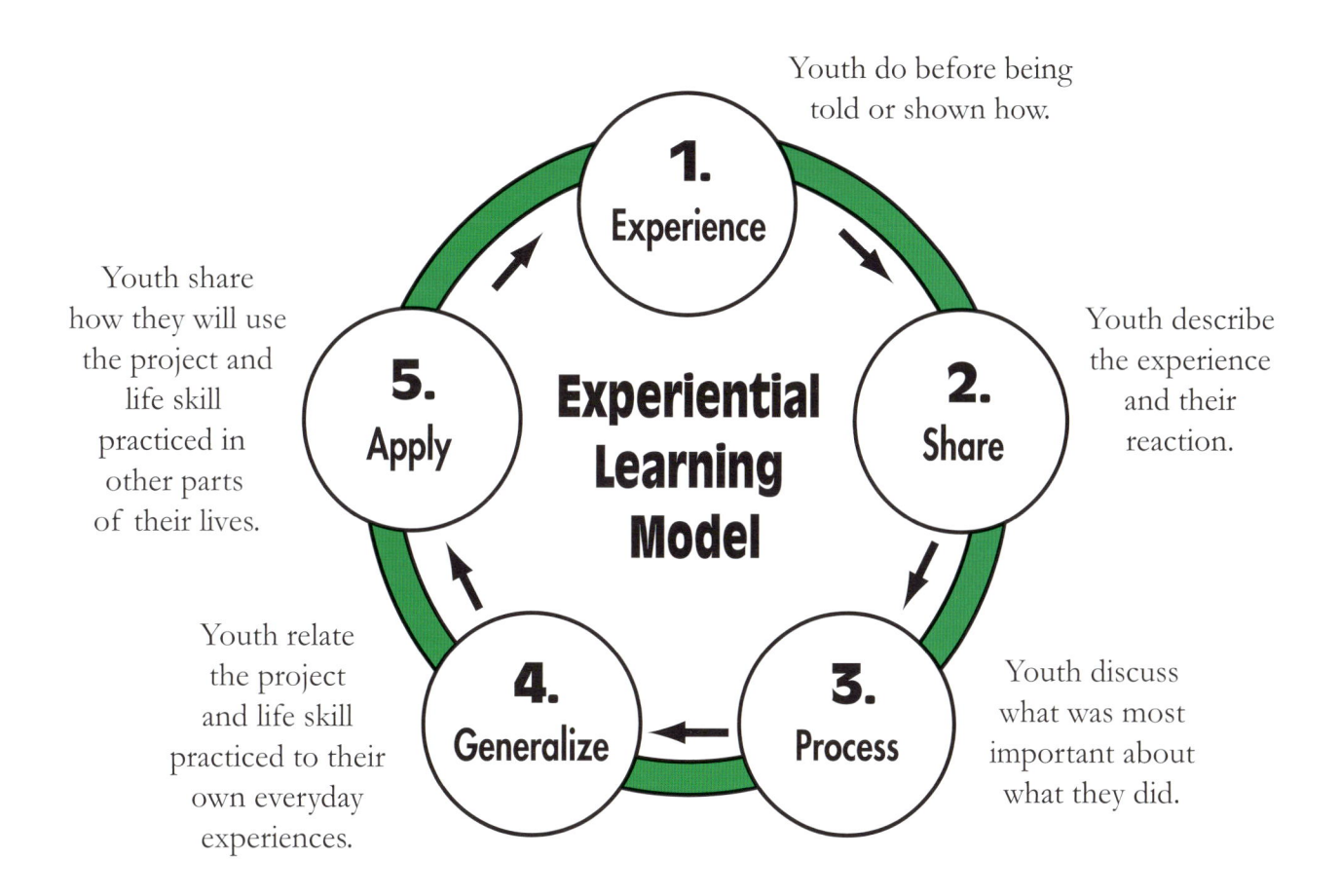

Experiential Learning Model

1. **Experience** — Youth do before being told or shown how.
2. **Share** — Youth describe the experience and their reaction.
3. **Process** — Youth discuss what was most important about what they did.
4. **Generalize** — Youth relate the project and life skill practiced to their own everyday experiences.
5. **Apply** — Youth share how they will use the project and life skill practiced in other parts of their lives.

Pfeiffer, J. W., & Jones, J. E., *Reference Guide to Handbooks and Annuals* 1983, John Wiley & Sons, Inc. Reprinted with permission from John Wiley and Sons, Inc.

Facilitation of the Experiential Learning Process

The key to the experiential learning process is that youth seek answers to questions rather than being given answers. This process requires facilitation instead of instruction. The role of adults in the experiential learning process is to facilitate the learning process, which means they become co-learners (Maxa, et. al., 2003).

Questioning Strategies

Questions suggested by the facilitator are designed to help promote discussion and engagement with life skills and science, engineering and technology abilities. The goal is to have the questions reside with the learner. Questions should promote discussion, interaction, and stimulate learner thinking. They should encourage ideas, speculation, and the formation of hypotheses. This type of questioning will not lead to a single right answer, but it will promote deeper understanding. It is important to allow adequate time for questions and discussions that engage youth and enhance learning. The process cannot be rushed.

Positive Youth Development

High quality 4-H Science programming provides valuable benefits in engaging youth. These programs also give youth the opportunity to engage in a positive youth development setting. Positive youth development occurs from an intentional process that promotes positive outcomes for young people by providing opportunities, choices, caring relationships, and the support necessary for youth to fully participate in families and communities. Simply, positive youth development seeks to develop young people as resources instead of problems to be managed (Lerner, 2005). Creating a positive youth development setting requires that youth are able to develop a sense of competence in social, academic, cognitive, health, and vocational aspects of life, a feeling of self-efficacy, a positive bond with a caring adult, a respect for societal and cultural norms, a sense of sympathy and empathy for others, and make contributions to self, family, communities, or society (Lerner, 2005).

4-H National Headquarters has identified four essential elements of positive youth development. They include:

- Belonging – To know they are cared about by others
- Mastery – To feel and believe they are capable and successful
- Independence – To know they are able to influence people and events
- Generosity – To practice helping others through their own generosity

To ensure that youth are engaged in a positive youth development setting, it is critical that learning includes the 4-H essential elements of positive youth development. This will give youth the opportunity to develop positive youth/adult relationships, practice life skills, and engage in the experiential learning model, which can promote mastery, independence, and generosity.

4-H Life Skills

Life skills are important in helping youth become self-directing, productive, contributing members of society (Maxa, et. al., 2003). 4-H programming strives to give youth developmentally appropriate opportunities to experience life skills, to practice them until they are mastered, and use these skills throughout a lifetime (Hendricks, 1998). The Targeting Life Skills model is based on the 4-H clover, which represents head, heart, hands, and health.

Targeting Life Skills Model

The "Targeting Life Skills Model" helps identify developmental life skills in 4-H and Youth Development Educational programs. They are grouped by:

Head Managing, Thinking
Heart Relating, Caring
Hands Giving, Working
Health Living, Being

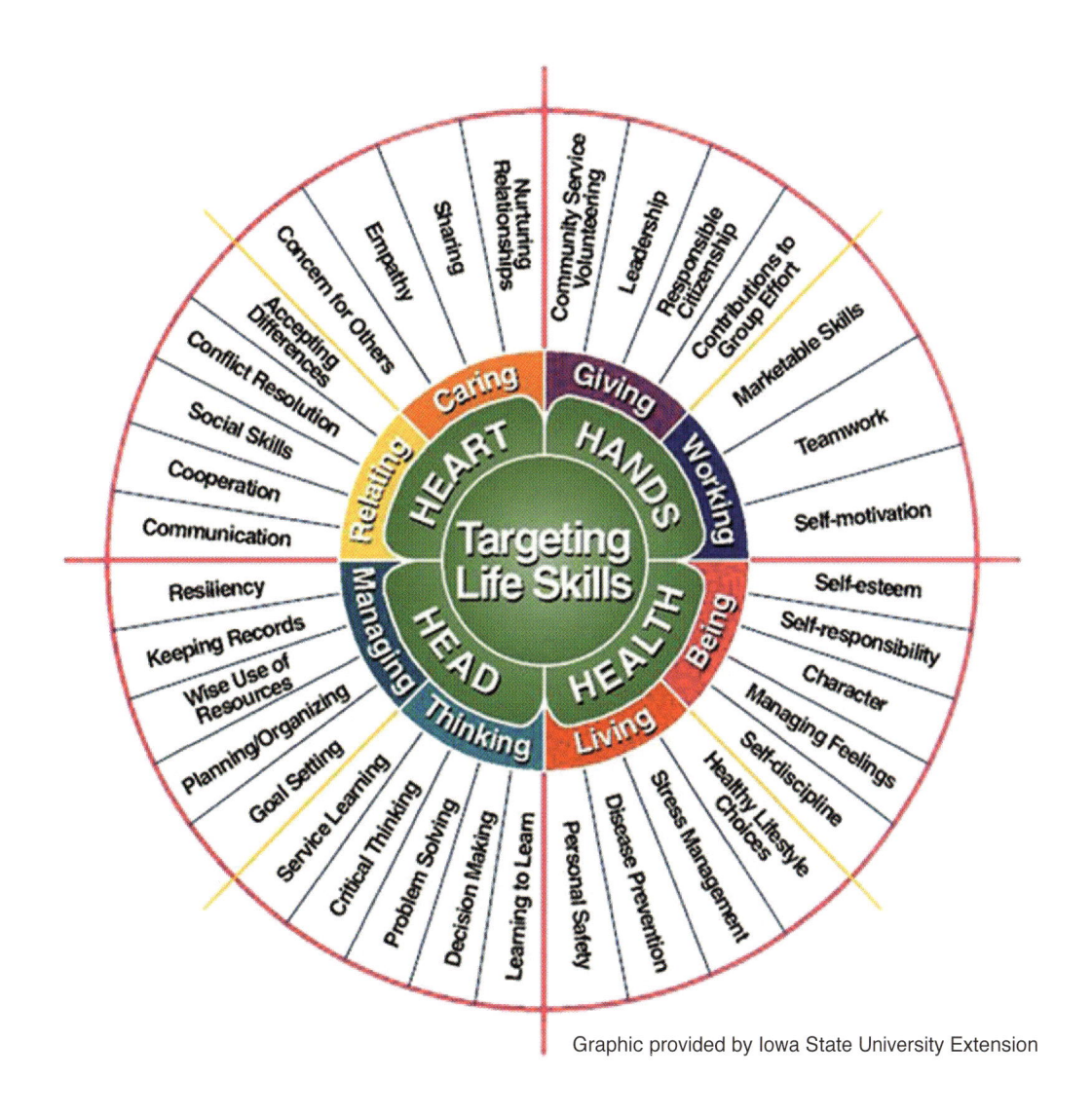

Graphic provided by Iowa State University Extension

What You Will Need for Junk Drawer Robotics

1. Trunk of Junk

The Trunk of Junk is a collection of items that will be used throughout this project. You may collect many of the items before you begin or add items as needed. Items you might collect are:

- Cardboard tubes from gift wrap, aluminum foil, paper towels, etc.
- Stationery supplies, like paper clips, binder clips, paper brads, etc.
- Clothespins, used household items, old toys, and items to take apart
- Coffee stirrers (sticks and straws), drinking straws, paper and plastic cups
- Construction paper, copy paper, graph paper, etc.
- Various pieces of cardboard; flats, boxes, tubes
- Assortment of small bolts, nuts, washers, and screws
- Wooden sticks (paint sticks, plant stakes, craft sticks, etc.) **Note**: Wooden sticks with holes are very useful in building. Holes can be predrilled or youth can drill holes as needed if you have appropriate tools. The website at *www.4-H.org/curriculum/robotics* includes tips on drilling holes in craft sticks.

2. Activity Supplies

Each activity has a list of the items to be prepared for that particular activity. Advance preparation or photocopying may be required. Some of the supplies may come from the Trunk of Junk, and some will be specific to the activities, but most will need to be organized before the meeting. Specific instructions are included in each activity.

3. Toolbox

This is a selection of basic hand tools that can be used throughout the project. These may be stored together in a toolbox, or collected before each meeting as needed. Commonly used tools include:

- Glue, tape, scissors (one or two per group)
- Low-temperature glue gun (two or three to share)
- Hand drill with different size bits
- Small hacksaw (to cut dowels and small boards)
- Pliers, wire cutters, screwdrivers, small hammer
- Bench hook or work surface to protect tables and chairs
- Paper towels, brush, cleanup supplies
- Punches (leather, craft, hole)

4. The Presenter's Activity Guide

This booklet is one of the three Presenter's Activity Guides: Give Robots a Hand, Robots on the Move, and Mechatronics. Each guide includes three to five modules made up of a series of activities. Three types of activities are within each module. In the To Learn activities, the focus is on gaining knowledge as a scientist. The To Do activities promote engineering design and innovation. And the To Make activities develop technological skills.

5. Robotics Notebook

There is one Robotics Notebook for the three levels of the *Junk Drawer Robotics* curriculum. Using the notebook encourages members to think like scientists and engineers. In their notebook, they will record their ideas, collect data, draw designs, and reflect on their experiences. It also provides specific information for the challenges. Each youth should have his or her own Robotics Notebook. If this is not possible, a blank notebook can be substituted. Graph paper will work best. Youth will have to record both the questions and their responses in a blank notebook, and leaders will have to make copies of supporting material.

6. Website and On-line Resources

The 4-H Robotics website, *www.4-H.org/curriculum/robotics*, provides supporting information, including background about robotics concepts, tips, and resources for leaders. You will find detailed tips on gathering tools, locating supplies, and instructions on how to make some of your own materials for *Junk Drawer Robotics*. There are also resources specifically for coaches or mentors working with a team of teens as they present the *Junk Drawer Robotics* modules. The role of a coach or mentor is to provide the teens with back-up support and resources. The website also has an overview of how the *Junk Drawer Robotics* activities are related to other parts of *4-H Robotics: Engineering for Today and Tomorrow* curriculum and suggestions for implementing the curricula in different settings like camps, events, clubs, or afterschool programs.

7. Additional 4-H Robotics Opportunities

Extend your experience with robotics with another part of the *4-H Robotics: Engineering for Today and Tomorrow* curriculum. *4-H Virtual Robotics* provides the opportunity to utilize an interactive computer game environment to learn about the science and engineering of robots. *4-H Robotics Platforms* provides a curriculum to use as you explore a robotics kit and learn to design, build, and program a robot. For more information or to order the curriculum, visit the website at *www.4-H.org/curriculum/robotics*.

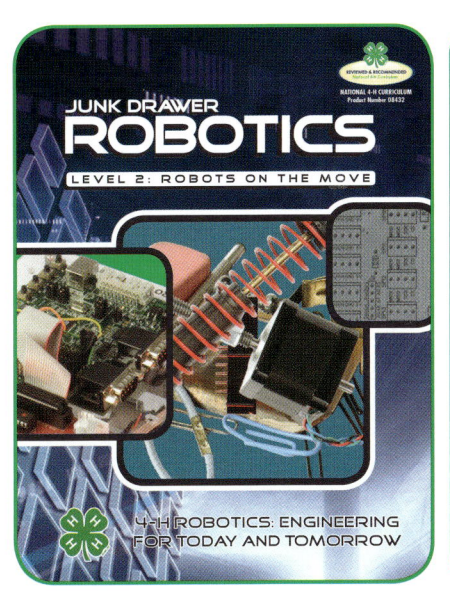

Focus for Robots on the Move

Introduction

Mobile robots make up an array of applications, allowing robots to go and do things in locations and situations that humans would not like or could not go. Mobile robots can go into dangerous situations, can fly, go under water, traverse remote areas or planets, carry instruments and sensors, and much more. In this level, *Robots on the Move*, we will explore legged, wheeled, and underwater robots. We will explore friction, basic electrical power and motors, engineering constraints, gear systems, and buoyancy.

Big Ideas

Friction can slow down or limit the movement of objects, but friction is also a useful tool when we need traction or gripping power.

Understanding of underlying physical science and mathematics concepts is necessary in making engineering design decisions.

Engineering design is a purposeful process of generating and evaluating ideas that seeks to develop and implement the best possible solution to a given problem.

Scientific habits of mind (observation, computation, communication, evaluation) are an important element of science literacy.

An electronic circuit is a closed path through which an electric charge can flow – typically consisting of a power source (battery), conductors (wire), a load (bulb), and control (on/off switch).

The interaction between the "sense," "think," and "act" components of what a robot does is accomplished using electric signals.

Mobile robots can be used on land, in the air, on the surface, or under water. A mobile robot can be controlled remotely using a cord or tether, radio control, or with programmed instructions.

A drive train directs the motion, speed, and direction of the movements from a motor. Gears, levers, and cams are components of a drive train.

Robots often use simple machines (such as levers, gears, or wheels and axles) as tools to accomplish their intended function.

Buoyancy is the force that causes an object to float in a fluid. An underwater ROV needs to be near neutral buoyancy so it can operate under water.

Connected Ideas

Robot Mobility is presented in
  Virtual Robotics: Robot Tractor Pull,
  Junk Drawer Robotics: Get Things Rolling,
  Junk Drawer Robotics: Get a Move On,
  Robotic Platforms: Build a Bot, and
  Robotic Platforms: Pick a Motion Challenge.

Electronic Systems/Circuits is presented in
  Virtual Robots: Electronics: Power Up!,
  Junk Drawer Robotics: Watt's Up?,
  Junk Drawer Robotics: Robots on the Move, and
  Junk Drawer Robotics: It's Logical.

Robot Mechanics/Simple Machines is presented in
  Virtual Robotics: Robot Mechanics,
  Virtual Robotics: Robot Tractor Pull,
  Junk Drawer Robotics: Give Robots a Hand,
  Junk Drawer Robotics: Get a Grip, and
  Robotic Platforms: Gears and Levers.

National Science Education Standards (NSES)

Unifying Concepts and Processes:
  Form and function
Science as Inquiry:
  Abilities necessary to do scientific inquiry
Physical Science Standards:
  Motion and forces
  Light, heat, electricity, and magnetism
Science and Technology:
  Abilities of technological design
  Understanding about science and technology

Standards for Technological Literacy (STL)

The Nature of Technology
  Scope of technology
  Core concepts of technology
  Relationships among technologies
Technology and Society
  Influence of technology
  Use of technology
Design
  Attributes of design
  Engineering design
  Problem solving
Abilities for a Technological World
  Apply the design process
  Use technological systems
  Impact of products and systems
The Designed World
  Energy and power technologies
  Transportation technologies
  Manufacturing technologies

Science, Engineering and Technology Abilities Developed in this Level

- Build/Construct
- Communicate/Demonstrate
- Collect Data
- Design Solutions
- Draw/Design
- Hypothesize
- Observe
- Predict
- Redesign
- Test
- Use Tools

Life Skills Practiced in this Level

- Communication
- Contributions to Group Effort
- Critical Thinking
- Keeping Records
- Problem Solving
- Sharing
- Teamwork

Module 1: Get Things Rolling

Overview of Activities in this Module

To Learn
Activity A — Slip N Slide
Activity B — Rolling Along

To Do
Activity C — Clipmobile Design Team

To Make
Activity D — Clipmobile Build Team

Note to Leader

When two people don't seem to get along, we say there is friction between them. What is friction? In physics, we might think of friction as a resistance to motion or movement. What affects friction? The type of surface — is it smooth or rough? Is it stationary or already moving? (If it is moving, the object has momentum.) The mass or weight of the object also can affect the amount of friction.

Try this: Rub your hands together. What do you feel? First, you only feel the surface of your palms and fingers; then they get warm. In rubbing, the surface particles change the movement or kinetic energy into thermal energy or heat that you feel in your hands.

Friction can slow down or limit the movement of objects, but friction is also a useful tool when we need traction or gripping power. What we need to find is the right amount of friction for the current use. Do we need wheels and gears that can turn freely on their axes? Do we need wheels that can grip the road to move a robot forward or up a hill? Sometimes we need to both reduce and increase friction.

Uses of friction in everyday life can be seen when we walk or ride in a car. Have you slipped in spilled water, or on ice? Have you seen a car spin around on ice because it "lost its friction"? These are examples of where we need friction.

On the other hand, friction can make it too hard to move or slide some objects. Friction also can cause a number of concerns that we should try to deal with. These include making it hard to move things by having to use more force or energy to overcome friction; losing some of our energy to heat; and ruining some objects because the heat generated by friction makes them wear out. Engineers try to make moving objects as efficient as possible; that is, they try to convert as much energy into target work as possible.

What can we do? If we need to reduce friction, we can try to use lubricants, rollers, or sliders to move objects. For more grip, we might try new tires, surfaces with grooves, rubber soles on shoes, and cushion grips on handles.

In addition to the physical constraints of friction, the Design and Build Teams will also consider customer requirement constraints such as carrying capacity, cost limits, complexity, and efficiency in material usage.

What you will need for Module 1: Get Things Rolling

- Robotics Notebook for each youth
- Trunk of Junk, see page 8
- Activity Supplies
  - Cardboard for ramp – at least one to share with the groups or one for each group
  - Paper clips, binder clips in different sizes, about 10-15 per group
  - Full boxes of paper clips (at least one per group)
  - Paper brads, about 5 to 10 per group
  - Clothespins or other fasteners, about 5-10 per group
  - Coffee stirrers, about 10-15 per group
  - Straws, about two to six per participant
  - Craft sticks (some with holes), about 10-15 per group
  - Toy wheels, at least four per group
  - Various kinds of tape (e.g., electrical tape, aluminum tape, duct tape, masking tape, and/or packing tape – 1 ½ to 2 inches wide, if possible)
  - Pieces of two or three different grits of sandpaper
- Toolbox
  - Scissors, a few to share with all
  - Small hacksaw (if needed)
  - Protractor to measure angles
  - Drill and bits

Timeline for Module 1: Get Things Rolling

Activity A – Slip N Slide
- Activity will take approximately 20 minutes.
- Divide youth into small groups of two or three.
- For each team, provide a box of paper clips and cardboard ramp with various surfaces attached.

Activity B – Rolling Along
- Activity will take approximately 20 minutes.
- Divide youth into small groups of two or three.
- Add to supplies from Activity A: paper clips, drinking straws, coffee stirrers.

Activity C – Clipmobile Design Team
- Activity will take approximately 20 minutes.
- Divide youth into groups of two or three.

Activity D – Clipmobile Build Team
- Activity will take approximately 30 minutes.
- Use the same groups from Activity C, Clipmobile Design Team.

Focus for Module 1: Get Things Rolling

Big Ideas
- Friction
- Underlying physical science and mathematics concepts
- Engineering Design Constraints
  - Complexity
  - Efficiency
  - Capacity
  - Cost/Budget

NSE Standards
- Form and function
- Motion and forces
- Abilities of technological design

Performance Tasks For Youth

You will explore movement and friction by testing a small box on a number of surfaces, looking at static friction and sliding friction.

You will test rolling friction by adding wheels (cylinders) or rollers as a way to overcome the overall friction of an item.

You will plan and design a vehicle to maximize its ability to coast based on considering the effects of friction. You will also consider constraints of capacity, efficiency, complexity, and costs in the design.

You will build or assemble a complex clipmobile, considering and addressing effects of friction and design constraints.

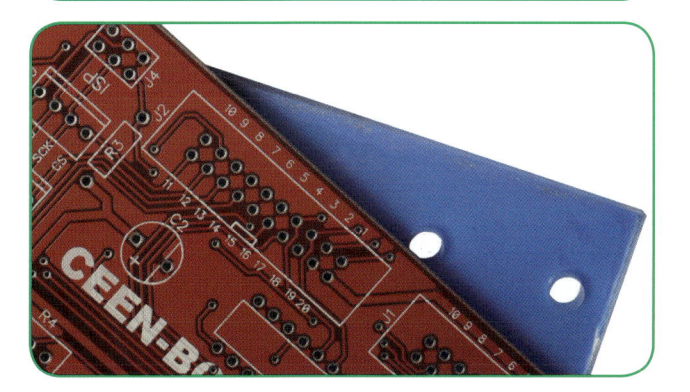

STL
- Apply the design process
- Use of technology
- Core concepts
- Transportation technologies
- Manufacturing technologies

SET Abilities
- Collect Data
- Draw/Design
- Hypothesize
- Observe
- Predict

Life Skills
- Critical Thinking
- Keeping Records
- Sharing

Success Indicators

Youth will be able to compare and select materials based on how they may affect sliding friction.

Youth will be able to demonstrate and discuss how rolling friction requires less force than sliding friction.

Youth will design a vehicle that will roll easily and meet the constraints listed.

Youth will build a vehicle to overcome friction and other constraints to roll freely down a ramp.

Activity A – Slip N Slide

Performance Task For Youth

You will explore movement and friction by testing a small box on a number of surfaces, looking at static friction and sliding friction.

Success Indicators

Youth will be able to compare and select materials based on how they may affect sliding friction.

List of Materials Needed

- Robotics Notebook
- Trunk of Junk
- Activity Supplies
  - Cardboard ramp, about 12 inches by 36 inches – at least one to share or one per group
  - Box of paper clips (two per group)
  - Paper clips – 6-10 per group
  - Pieces of various tape (e.g., masking tape, packing tape, aluminum tape, duct tape), each piece about 10 inches long to attach to ramp
  - Sandpaper strips about 10 inches long — in fine, medium, and coarse grits — to attach to ramps
  - Protractor to measure ramp angle

Activity Timeline and Getting Ready

- Activity will take approximately 20 minutes.
- Divide youth into groups of two or three.
- Before the meeting, fasten strips of different types of tapes and/or sandpapers on the surface of the cardboard ramp so that items can slide down as the ramp is raised at one end. Depending on the size of the cardboard, you may be able to get four or five different surfaces on a ramp. On each ramp, leave a blank, uncovered strip of just the cardboard for a control run. Depending on the variety of tapes you have, you may place different types of tape or sandpaper on different cardboard ramps.
- Provide teams with a cardboard ramp, protractor, and box of paper clips.

Experiencing

1. Share with youth the difference in static friction of an object at rest and that of sliding friction when it starts to move. Tell youth that they will be testing some objects to see how these starting sliding friction points can be different for a number of reasons. Ask the group if they have any ideas on how to measure the angle of the ramps. Share with them that they can use the protractors and have them figure out and practice measuring the ramp angles.

2. Youth will begin by testing their control (plain cardboard). Youth will place the box at the top of the plain cardboard "track." The youth will slowly raise the cardboard until the box begins to slide down the track. They will then record this angle and repeat the experiment two more times. This will be the control test, and the angle used for comparison. This is a good time to have the youth predict if they think certain materials will allow the box to start sliding sooner (less of an angle) or not start sliding until at a higher angle.

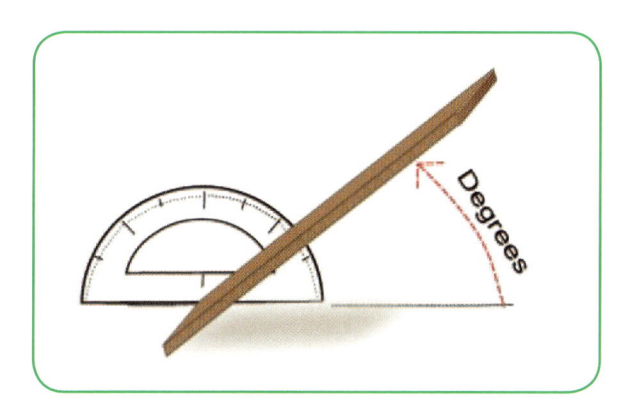

3. Youth will then test a different material on their cardboard ramp, comparing these angles to the angle of the control test.

Sharing and Processing

As the facilitator, help guide youth as they question, share, and compare their observations. Before they share with the group, have youth reflect on the activity in their Robotics Notebook. Use more targeted questions as prompts to get to particular points. There is no one right answer.

- Ask each group to prepare a summary of its results and form a hypothesis about why changing the surface area had an effect on how the box would slide. Ask each group to share with everyone.
- If the data differed among the groups, discuss why that might be.
- What had the greatest effect on the sliding friction?
- What had the least effect on the sliding friction?
- Why do you think it was important to repeat the same test more than once? How many times should a test be repeated?

Generalizing and Applying

- If you wanted to have more grip or traction (more friction), what type of tape or material could you use?
- If you wanted to reduce the friction, what type of tape or material could you use?
- Youth can apply what they have designed in Activity B.

Activity B – Rolling Along

Performance Task For Youth

You will test rolling friction by adding wheels (cylinders) or rollers as a way to overcome the overall friction of an item.

Success Indicators

Youth will be able to demonstrate and discuss how rolling friction requires less force than sliding friction.

List of Materials Needed

- Robotics Notebook
- Trunk of Junk
- Activity supplies — same as for Activity A plus
  – Coffee stirrers (straws)
  – Drinking straws, different-sized diameters, if possible – two to four per group
  – Paper clips — six to eight per group
  – Cardboard ramp with different surfaces to test – one per group or some to share

Activity Timeline and Getting Ready

- Activity will take approximately 20 minutes.
- Divide youth into groups of two or three.
- Provide each team with supplies.

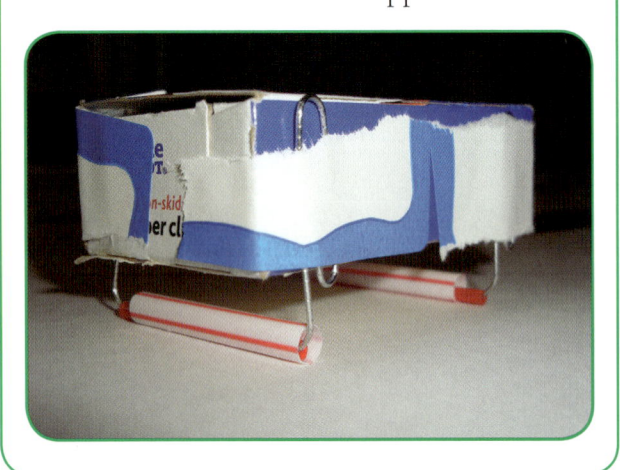

Experiencing

1. Share with the group that in the last activity they were exploring sliding friction, and in this activity they will be looking at rolling friction. Ask if anyone can describe what the difference might be. There are no right or wrong answers — have the youth discover through active discussion. Generally, rolling friction is with wheels or cylinders and with spheres or balls. Each of these items will have much smaller touch areas between the surfaces and that — in part — plus other factors will reduce the friction. Share that in the next tests they will be using rollers to try to get the box to move more freely down the ramp.

2. Members will create axles and cylinder wheels using paper clips and pieces of drinking straws and/or coffee stirrer straws. The members will bend the paper clips so part of the paper clips can be mounted to the box with tape and part can hang down for an axle. Using short pieces of straws, slide them onto the paper clip axles. If there are different sizes of straws, members might want to try each one or try putting one straw inside another. They can try one or more axles.

3. Members will test the effect of rollers and wheels. Members will place the box of paper clips on the control (plain — no tape or sandpaper) part of the cardboard, and slowly raise the cardboard until the box of paper clips begins to roll down. They will then measure the angle at which the box of paper clips begins to move.

4. Members will then test the rollers and wheels on different types of surfaces, repeating the same steps as above.
5. Members should repeat each test three times making sure their readings are accurate.

Sharing and Processing

As the facilitator, help guide youth as they question, share, and compare their observations. Before they share with the group, have youth reflect on the activity in their Robotics Notebook. Use more targeted questions as prompts to get to particular points. There is no one right answer.
- Ask the groups to prepare a summary of their results and form a hypothesis about why the boxes rolled better or worse than in the first activity without the rollers. Ask each group to share with everyone.
- How did rolling make a difference on the different surfaces versus sliding on the surfaces?
- Share your experience in making axles and wheels (cylinders).
- What do you think would improve the axles and wheels?

Generalizing and Applying

- Where have you seen or heard about using rollers to move heavy loads?
- Youth can apply what they have learned in Activity C.

Activity C – Clipmobile Design Team

Performance Task For Youth

You will plan and design a vehicle to maximize its ability to coast, based on considering the effects of friction. You also will consider constraints of capacity, efficiency, complexity, and costs in the design.

Success Indicator

Youth will be able to design a vehicle that will roll easily and meet the constraints listed.

List of Materials Needed

- Robotics Notebook
- Activity Supplies
  - A bag of "start-up" sample supplies for each Design Team. One each of the following is suggested:
    - One regular craft stick, one jumbo craft stick, one craft stick with holes
    - One regular paper clip, one large paper clip
    - One 1-inch paper brad, one 1 ½-inch paper brad
    - One binder clip
    - One drinking straw
    - One coffee stirrer (straw type)
    - One rubber band
    - One toy wheel
    - One wood skewer (dull or remove the sharp, pointed end; may cut or break in half for packing in bag)
- Copies of play money on page 26
- Optional materials could include whiteboard, poster pad, or newsprint.
- Optional: Poster/Handout of Clipmobile Challenge on page 24
- Optional: Copies of Junk Drawer Supply Company Materials Order Form on page 25

Activity Timeline and Getting Ready

- Activity will take approximately 20 minutes.
- Divide youth into groups of two or three.
- Assemble packs of start-up materials for each group.
  - Fill a resealable bag with some materials that can be used in building the Clipmobile — one bag for each group.
  - Each bag should include some of each type of supplies available to the groups, but the bag should not include enough supplies to complete a full Clipmobile.

Experiencing

1. Explain to youth about the budget. They will have $45.00 to spend on materials that they will "purchase" from the materials supplier (you). Youth must learn the importance of staying on budget. For every dollar ($1) under budget, they get an extra point. For every dollar ($1) over budget, they must subtract two points. In determining budget, youth cannot return any materials purchased, but get dollar points for supplies in inventory at half the value of new ones. Materials broken or unusable will not have any credit value.

2. Review with the groups the concepts of design and manufacturing elements, and of form and function (from module 1 in *Give Robots a Hand*).

3. Highlight that engineers work at solving problems based on constraints in the task's requirements. Some of those constraints will be listed in the Clipmobile activity.

4. Give limited instructions on the design task and customer requirements (constraints).

    a. Provide each team with their start-up funds ($45.00 play money).

    b. Ask each team to review the Clipmobile Challenge and the Junk Drawer Supply Company procedures (found in the Robotics Notebook or on page 24).

    c. Clipmobile Challenge (customer requirements):

        i. Design a vehicle that will roll freely down a ramp and will travel (coast) a long distance (overcome friction).
        ii. It must be able to hold (carry) a box of paper clips (capacity).
        iii. It must contain at least five different types of parts (complexity).
        iv. But it must use the least total number of all parts (efficiency).
        v. Cost target: under $35 of play money for all the parts and materials used to build one of their clipmobiles as designed.

5. Design Teams may purchase a sample bag of materials at the reduced cost of $10.00 for planning and design.

6. Allow Design Team discussion and design on how to make the vehicle. Teams may use start-up supplies during design, but this is not a build time so no tools or building during this time.

7. Have each Design Team use the Robotics Notebook to plan and draw their ideas. Groups should make sketches of their plans.

8. Based on their design plan, each team will create a Materials Order Form (MOF) of all the supplies they wish to order (purchase) for building their Clipmobile.

Sharing and Processing

As the facilitator, help guide youth as they question, share, and compare their observations. Before they share with the group, have youth reflect on the activity in their Robotics Notebook. Use more targeted questions as prompts to get to particular points. There is no one right answer.

- Ask Design Teams to discuss the vehicles they have designed with the entire group.
- How do you think friction will affect this vehicle?
- How did you use axles or bearings in the vehicle?
- Based on the design criteria, which part of the vehicle design is most important?
- What design criteria has to be included?
- How did you address all of the constraints of the customer requirements?
- What other parts or supplies might make it easier to design this vehicle?

Generalizing and Applying

- Where might manufacturers have to be concerned about friction when they design a car?
- How about in designing a farm tractor or plow?
- Select an item either in the room or one commonly found and discuss constraints that might have been used during its design.
- Youth can apply what they have designed in Activity D.

CAREER CONNECTIONS

Career Connection 3: Constraints

There are many elements for engineers to consider when approaching a problem. Oftentimes, there are constraints that impact the engineering design process, especially in the type of design that can realistically be built. Engineers use their creativity and resources to overcome these obstacles.

Engineers must take time into account when devising a solution, because they often work on a deadline. "Time to market" is used to describe the time needed to plan, create, test, produce, and release a new product. The timely delivery of products to people is critical for companies to profit. To minimize their "time to market," engineers work in teams, sometimes assigned to specialized components of the overall product.

Money is another constraint that engineers face. Engineers constantly search for less expensive materials that perform similarly to their expensive counterparts. Along with the expense of resources, engineers also must consider the availability of these materials. Certain supplies are sometimes unavailable, so engineers must find replacements and alternative equipment.

The physical elements of components also pose a challenge. Engineers must use physics and mathematics to ensure that the dimensions of their device will allow it to function correctly. They must find materials with properties that best suit the design.

- What are some constraints that you face in daily life? What are some ways you have found to work within those constraints?
- What do you think are some of the constraints in building a robot? Why?

Clipmobile Challenge

Using what you have learned by exploring friction and movement, see how you can apply your knowledge for this design challenge.

Customer Requirements for a Clipmobile

- Design a vehicle that will overcome friction and roll freely down a ramp, and travel a long distance (performance).
- It must be able to hold (carry) a box of paper clips (capacity).
- It must contain at least five different types of parts (complexity).
- But it must use the least total number of all parts (efficiency).
- Cost target is to be no more than $35.00 of play money, including start-up supplies in inventory bag (budget).

Manufacturing Selection Criteria

Criteria to select the team that will be chosen to mass produce the Clipmobile:

Meeting Design Criteria (constraints)

Capacity – carry a box of paper clips: (Yes) 10 points; (No) 0 points _____

Performance – roll down ramp and coast: +1 point per inch – maximum 24 points _____

Complexity – various types of parts used: +2 points for each type of part _____

Efficiency – least overall number of parts: -1 point for each part used _____

Budget/cost – cost of production: +1 point for each dollar under $35.00 _____

-2 points for each dollar over $35.00 _____

Team Business Strength

Capital – dollars left from $45.00   +1 point for each dollar still in cash _____

Inventory value – value of supplies in inventory   +1 point for each dollar of value _____

Overall Team Score _____

Junk Drawer Supply Company

JDSC is the official supplier of all materials for Clipmobile design, development, and manufacturing. Thank you for using JDSC.

How to get parts and supplies:

1. The Junk Drawer Supply Company will provide each Design Team with a Materials Order Form (MOF) and a sample of the different items for sale. This sample pack of supplies will only cost $10.00 for the whole bag of supplies (Over a $25.00 value of supplies — what a bargain!). These items can be used in your design process and in your building activity.
2. During the *Design*, complete the MOF, listing the total number of each part or item you plan to use.
3. Calculate the cost for each type of part you have ordered.
4. Add up all the costs for parts to get the total dollar amount needed to purchase all the items.
5. During the *Build activity*, have one of your team members take the completed MOF to the Junk Drawer Supply Company area to pay for and pick up the materials ordered.
6. The Junk Drawer Supply Company has limited operating hours and will close after all the teams have filled their orders. Make sure to order enough to build your Clipmobile, but try not to have too many extras as leftover inventory is only worth half of its cost new.
7. The Junk Drawer Supply Company is very picky and will not accept any returned parts or items.

Junk Drawer Supply Company
Clipmobile Materials Order Form (MOF)

Sold to: _____ Order Date: _____

Item Code #	Item/Part Description	Price per Item	Number Ordered	Total Cost (Price X Number)
101	**Craft Stick** – Large or small	$3.00		
102	**Craft Stick** *w/ holes* – Large or small	$4.00		
203	**Paper Clip** – Large or small	$1.00		
304	**Brass Paper Brad** – Various sizes	$1.00		
405	**Binder Clip** – Various sizes	$2.00		
506	**Drinking Straw** – Various sizes	$2.00		
507	**Coffee Stirrer Straw**	$1.00		
608	**Rubber Band** – Various sizes	$1.00		
709	**Wheel** – Various sizes	$3.00		
810	**Wood Skewer** – Various sizes	$2.00		
Thanks for using **Junk Draw Supply Company.** See us first for all your robot supplies!			**Grand Total:**	

4-H Junk Drawer Robotics: Robots on the Move • Module 1: **Get Things Rolling**

Activity D – Clipmobile Build Team

Performance Task For Youth

You will build or assemble a complex Clipmobile, considering and addressing effects of friction and design constraints.

Success Indicator

Youth will be able to build a vehicle to overcome friction and other constraints to roll freely down a ramp.

List of Materials Needed

- Robotics Notebook
- Activity Supplies
    - Collection of parts or materials that can be used in creating the Clipmobile
    - Different sizes of paper clips, binder clips, paper brads, and clothespins
    - Craft, paint, or wood sticks
    - Various sizes of coffee stirrers and drinking straws (axles and bearings)
    - Variety of other items, including toy wheels or disks, at least four per group
    - Cardboard for ramp – about 12 inches by 36 inches in size; one or more ramps can be shared for test runs.
    - Optional: Copies of Cost of Production and Materials Inventory sheets on page 30
- Toolbox
    - No tape or glue to be used in making the Clipmobile
    - Hand drill and bits
    - Hacksaw
    - Pliers, scissors, punches, if needed

Activity Timeline and Getting Ready

- Activity will take approximately 30 minutes.
- Use the same teams from Activity C.

Experiencing

1. Review design challenge and customer requirements for the Build Teams.
    a. Build a vehicle that will roll freely down a ramp and will travel (coast) a long distance (overcome friction).
    b. It must be able to hold (carry) a box of paper clips (capacity).
    c. It must contain at least five different types of parts (complexity).
    d. But it must use the least total number of all parts (efficiency).
    e. Cost target is to be less than $35.00 of play money, including start-up supplies in inventory bag (budget).
2. Build Teams will start with the following from the Design Team activity:
    a. Start-up bag of sample supplies used in design activity
    b. $35.00 of play money start-up funds (dollars left after purchase of start-up supply bag)
    c. Materials Order Form completed in the design activity
    d. Clipmobile plans created in the design activity
3. Allow each team to go to the Junk Drawer Supply Company (JDSC) with their MOF (Materials Order Form) and play money to purchase supplies.
    a. The materials for the Build Teams will need to be organized for "sale" to the teams.

b. The parts may be sorted and stored in plastic bags or other containers for easy distribution.

c. The JDSC will need play money for change when selling items.

d. The JDSC will be open at the beginning of the build activity and parts will be distributed based on each Design Team's Materials Order Form.

e. The Build Teams will pay for all the items ordered and picked up.

f. The JDSC will close after filling the Materials Order Forms; no additional parts can be purchased or used.

4. Build Teams are to construct, test, and modify their Clipmobile designs.

5. When vehicles are complete, have the Build Teams complete their budget cost sheet and leftover supplies (inventory).

6. The teams are to present their vehicles, cost sheets, and inventory sheets, and demonstrate how their Clipmobile can hold the weight of a box of paper clips and roll down a ramp.

7. Youth should make notes in their Robotics Notebook to record actions and modifications (what worked, what didn't work, how it was modified).

Sharing and Processing

As the facilitator, help guide youth as they question, share, and compare their observations. Before they share with the group, have youth reflect on the activity in their Robotics Notebook. Use more targeted questions as prompts to get to particular points. There is no one right answer.

- What design shapes worked well?
- What design shapes did not work well?
- What are some ways these vehicles overcame or used friction to help in their design?
- What seemed to work well for axles?
- What would have to be changed to haul heavier loads?
- How did the customer constraints affect the building of the Clipmobiles?

Generalizing and Applying

- How would you get a bicycle to coast farther without peddling?
- Where on a bicycle would you want to have friction? Or less friction?
- Have groups review the data they have gathered in their Robotics Notebook from the previous activities. Ask them to develop a hypothesis about friction – what are the properties of friction?
- What are some additional constraints that you think could be placed on this build?
- Youth can apply what they have learned in Module 2, Watt's Up.

Build Team

Clipmobile Reports

How to figure **Cost of Production** of a Clipmobile:

1. Take your completed Clipmobile and evaluate all the components and parts used.
2. Use the COP (Cost of Production) sheet and fill in the information.
   - List the number of each type of part used; how many wheels, how many brads, etc.
   - Calculate the cost of each type of part used in the Clipmobile.
   - Check the last column for the types of parts you used in building the Clipmobile.
3. At the bottom of the COP, calculate the totals for:
   - The total number of parts
   - The total cost of the parts used to build your Clipmobile
   - The total of the different types of parts used

You may have some parts that you did not use or parts that you cut or bent, or that broke when working on them. Sort these extra leftover parts into two stacks: one stack is good parts that could be used on a different activity, and a second stack is the broken, bent, or used parts that are not like new. Since the first stack is good as new, take an inventory (count) of them and record it on the Materials Inventory Sheet (MIS).

How to determine **Supply Inventory** on hand after building the Clipmobile:

1. Take the stack of good unused parts:
   - Group them by their type of part.
   - Count how many of each type and record the number or the MIS (Materials Inventory Sheet) in the "number on hand" column.
   - Calculate the value of each type of part. (Notice that those which are "used" have a value less than "new" parts.)
2. On the bottom, add up the totals for:
   - Total number of parts on hand (in inventory)
   - Total value of parts in inventory
3. Have a representative of JDSC (facilitator or leader) verify your inventory stock and your MIS sheet.

Clipmobile Report – Cost of Production (COP)

Clipmobile Report Cost of Production (COP)					
Date of Production:		Location of Mfg.:		Manufactured by:	
Item Code #	Item/Part Description	Price Per Item	Number Used in Vehicle*	Total Cost (Price X Number)	Check Items Used in This Build
101	Craft Stick – Large or small	$3.00			
102	Craft Stick *w/holes* – Large or small	$4.00			
203	Paper Clip – Large or small	$1.00			
304	Brass Paper Brad – Various sizes	$1.00			
405	Binder Clip – Various sizes	$2.00			
506	Drinking Straw – Various sizes	$2.00			
507	Coffee Stirrer Straw	$1.00			
608	Rubber Band Various sizes	$1.00			
709	Wheel – Various sizes	$3.00			
810	Wood Skewer – Various sizes	$2.00			
*include full value even if only part of an item was used, cut in half, taken apart, etc.		Totals			
			Total Parts Used	Total Cost of Production	Total Parts Used

Clipmobile Report – Materials Inventory Sheet

Clipmobile Report Materials Inventory Sheet (MIS) List of supplies left over and in good condition				For office use only	
Date of Inventory:		Location of Mfg.:		Name of Manufacturer:	
Item Code #	Item/Part Description	Used Value Per Item	Number of Good Items Still on Hand*	Total Value (Price X Number)	Verification of Inventory on Hand
101	Craft Stick – Large or small	$1.50			
102	Craft Stick *w/holes* – Large or small	$2.00			
203	Paper Clip – Large or small	$0.50			
304	Brass Paper Brad – Various sizes	$0.50			
405	Binder Clip – Various sizes	$1.00			
506	Drinking Straw – Various sizes	$1.00			
507	Coffee Stirrer Straw	$0.50			
608	Rubber Band Various sizes	$0.50			
709	Wheel – Various sizes	$1.50			
810	Wood Skewer – Various sizes	$1.00			
*only include complete items in good usable condition; not those cut, drilled, bent, taken apart, etc.		Totals			
			Total Parts Not Used	Total Value of Parts on Hand (Inventory)	

Module 2: Watt's Up?

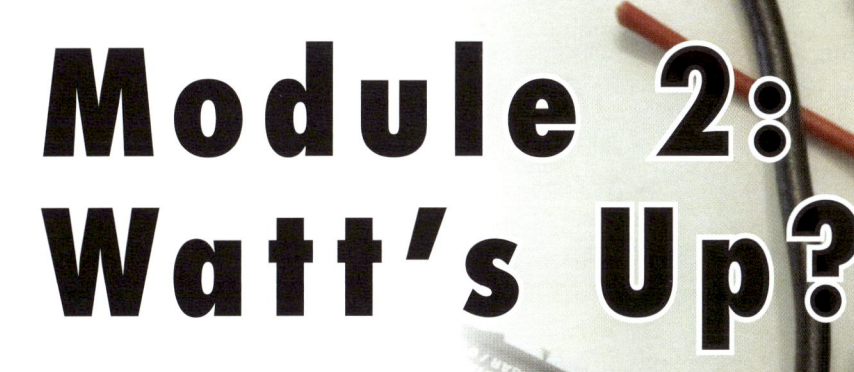

Overview of Activities in this Module

To Learn
Activity E – Light Up My Life
Activity F – Magnetic North

To Do
Activity G – Can-Can Robot Design Team

To Make
Activity H – Can-Can Robot Build Team

Note to Leader

In this module, youth will explore the use of electrical power using small toy motors, batteries, and a flashlight or holiday light bulbs. Participants will construct simple electrical circuits to run a motor and light up the bulbs. Standard AA, AAA, C, and D batteries produce about 1.5 volts of electrical power. This voltage is safe for participants as they explore in these activities.

An electrical flow or path is needed from the + (plus) battery contact through the item (bulb, motor, light, hair dryer, etc.) and back to the – (negative) contact of the battery. When there is a complete path, the electricity will flow and the light will glow, the motor will turn, the toaster will get hot, etc.

If participants wrap foil or wire around the battery ends and hold it in place, a short circuit is created and the wire may become hot. Should this happen, you can use follow-up questions about energy from batteries (electricity). Some uses of electric energy include heat, light, and magnetic force as seen in heaters, hair dryers, light bulbs, TV screens, motors, and electric solenoids. In these activities, youth will experience the power of heat from a battery, see light, and feel the movement from magnetic force.

Flashlight bulbs come in varying voltages (1.5, 3, 6, or more volts) depending on the type of flashlight (a flashlight with two batteries in series = 3 volts, with 4 batteries = 6 volts). Using a 6-volt bulb with one battery will be dimmer. A 1.5-volt bulb connected to five batteries (7.5 volts) may burn out quickly. Although the bulbs are relatively expensive to replace, this can be a learning experience if you question why the bulb burned out so quickly. An inexpensive lamp source is a mini holiday light bulb. Many are low-voltage bulbs with cords that can be cut to hold the lamp, making the wire connections easily.

Safety Note:

Glass bulbs may break, exposing ends of wire and sharp edges. The holiday bulbs will be used with batteries in these activities and not with household outlets of 110 volts. DO NOT attempt to use this higher voltage for any of the *Junk Drawer Robotics* activities.

In the activities from *Give Robots a Hand*, youth explored air power or pneumatics as a power source in robots. In these activities, youth will use a battery to build an electrical-powered cup robot to draw on paper.

What You Will Need for Module 2: Watt's Up?

- Robotics Notebook
- Trunk of Junk, see page 8
- Activity Supplies
  - Batteries (AA and C or D), at least one of each size per group
  - Aluminum foil cut or torn into strips about 6 inches long and 1 ½ inches wide
  - Flashlight bulbs
  - Rubber bands
  - Small toy motors (1.5 to 6 volts range) with 4- to 8-inch wire leads attached, one for each group
  - Two to four pieces of small wire (inside telephone wire or 18-22 gauge insulated copper wire) in 6- to 10-inch lengths. If the wire ends are not stripped, provide several pairs of wire strippers.
  - Thin felt tip markers of various colors of washable ink, three or four per group
  - Pencil top erasers, one per group
  - Paper or plastic cups, 8-ounce to 16-ounce Note: Some plastic cups may be harder for the participants to work with, such as in drilling or poking holes.
  - Butcher paper, newsprint, or poster paper for the robots to draw their designs
- Tool Box
  - Scissors, 1 or 2 per group
  - Masking tape
  - Electrical tape
  - Hand drill with bits
  - Small hacksaw (to cut dowels and small boards)
  - Soldering iron and electrical solder
  - Pliers, scissors, wire cutters/strippers
- Things to Make or Acquire
  - Soldering the lead wires onto the motor contacts and using electrical tape to fasten them to the side of the motor will help members construct their robots. See the 4-H Robotics website at *www.4-H.org/curriculum/robotics* for examples and safety tips.

Safety Tip:
Review soldering iron user's manual and safety instructions. Always wear leather or heat-resistant gloves and safety glasses when using a soldering iron.

Timeline For Module 2 Watt's Up?

Activity E – Light Up My Life
- Activity will take approximately 20 minutes.
- Make sets of one battery, one flashlight bulb, and two strips of foil for each group. Put each set into a plastic sandwich bag to keep the parts organized.
- Divide youth into groups of two or three.

Activity F – Magnetic North
- Activity will take approximately 20 minutes.
- Divide youth into groups of two or three.
- Provide each team with supplies.

Activity G – Can-Can Robot Design Team
- Activity will take approximately 20 minutes.
- Divide youth into groups of two or three.

Activity H – Can-Can Robot Build Team
- Activity will take approximately 30 minutes.
- Use the same groups from Activity G, Can-Can Robot Design Team.

Focus for Module 2: Watt's Up?

Big Ideas
- Science habits of mind
- Electric signals
- Electrical circuits
- Engineering design

NSE Standards
- Light, heat, electricity, and magnetism
- Abilities of technological design
- Understanding of science and technology

STL
- Relationships among technologies
- Influence of technology on history
- Use technological systems
- Problem solving
- Energy and power technologies

Performance Tasks For Youth

You will learn about electrical circuits by creating a circuit, causing a light to glow.

You will concentrate the electromagnetic field to deflect a compass needle.

You will design and sketch an electric-motor-powered robot made from a cup. The robot will be able to draw or make marks on a piece of paper.

Based on your design and using a cup, you will build a robot that will draw on paper.

SET Abilities
- Collect Data
- Draw/Design
- Hypothesize
- Observe
- Predict

Life Skills
- Communication
- Contributions to Group Effort
- Critical Thinking
- Keeping Records
- Sharing

Success Indicators

Youth will make connections on a battery so that an electrical circuit will light a bulb.

Youth will wind a wire coil and place it close to a compass to deflect the needle when current is flowing in the coil.

Youth will design and sketch an electric-motor-powered robot made from a cup. The robot will be able to draw or make marks on a piece of paper.

Using a cup, other materials, and their design, youth will build a robot that will draw on paper.

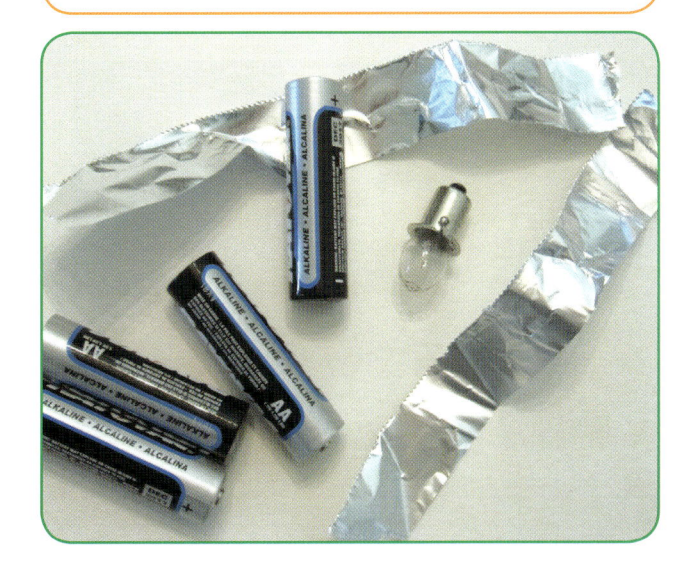

Activity E – Light Up My Life

Performance Task For Youth
You will learn about a simple electrical circuit by using batteries, foil, and light bulbs.

Success Indicators
Youth will be able to demonstrate an understanding of electrical circuits by lighting a bulb.

List of Materials Needed
- Robotics Notebook
- Activity Supplies
  - C or D batteries, one per group
  - Aluminum foil cut or torn into strips about 6 inches long and 1 ½ inches wide, two per group
  - Flashlight bulbs, one per group but have extras available
  - Masking tape
  - Optional: poster paper or whiteboard for sharing ideas

Activity Timeline and Getting Ready
- Activity will take approximately 20 minutes.
- Make sets of one battery, one flashlight bulb, and two strips of foil for each group. Keep each set in a plastic sandwich bag so it stays organized.
- Divide youth into small groups of two or three.

Experiencing

1. Challenge the youth to experiment with their materials to light the bulb.
2. When the task is complete, ask participants to draw a diagram in their Robotics Notebook showing the connections of the battery, bulb, and foil. Participants also should draw connections that didn't light up the bulb. Ask members to draw the connections of the battery, bulb, and foil on poster paper or a blackboard for others to see.
3. If successful with one bulb, have participants try to get two or three bulbs to light using just one battery.
4. Challenge the groups to connect the bulbs to the batteries so they light at different levels of brightness.

   Note: Multiple bulbs will glow differently in series/parallel circuits. Series and parallel circuits will be discussed in Level 3. Use one battery to avoid burning out too many light bulbs.

5. Challenge youth to carefully try to get the wire or battery hot. Ask youth to record in their Robotics Notebook why they think the temperature is rising. They can make a short circuit by a direct connection with the two ends of the battery. Prepare them to be ready to let go quickly when they touch both the positive and negative terminals of the battery with the foil at the same time. It should slowly get hot from the flow of electrons. When it does, they should let go of the foil and contacts.

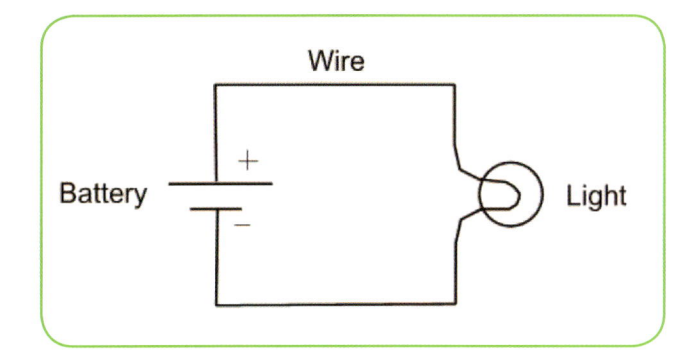

Diagram: A simple circuit showing a Battery (with + and − terminals) connected by Wire to a Light.

Sharing and Processing

As the facilitator, help guide youth as they question, share, and compare their observations. Before they share with the group, have youth reflect on the activity in their Robotics Notebook. Use more targeted questions as prompts to get to particular points. There is no one right answer.

- Ask youth to share how they got the bulb to light. Have them show what they drew in their Robotics Notebook or redraw it on a poster or blackboard.
- What would help make the bulb turn on and off easier?
- How was the circuit similar or different when the wire or battery was getting hot versus when it was lighting the bulb?

Generalizing and Applying

- How can an electrical circuit be used in a house? How about in a robot?
- What are some of the dangers of a short circuit?
- Describe the circuit in some common electrical items, such as a desk lamp, a toaster, a hair dryer, a blender, or others.
- Youth can apply what they have learned in Activity F.

Activity F – Magnetic North

Performance Task For Youth

You will learn about electromagnetic force in electrical circuits by using batteries, wire, and a compass.

Success Indicators

Youth will successfully use a coil of wire to move a compass needle.

List of Materials Needed

- Robotics Notebook
- Activity supplies
  – C or D batteries, one per group (can be the same as used in Activity E)
  – About 4 to 5 feet of thin wire (about 20 gauge) per group, similar to the small wire inside a phone cable. Wire needs to be insulated either with plastic cover or painted with varnish
  – Large steel nail (12p to 16p) or 1/4 inch by 3 inch bolt, one for each group
  – Magnetic needle compass, at least one to share in testing or enough for each group
  – Small magnet for each group
  – Paper clips and/or other small steel items to pick up, three to five per group
- Toolbox
  – Short pieces of masking or electrical tape
  – Wire strippers or side cutters
  – Needle nose or slip joint pliers

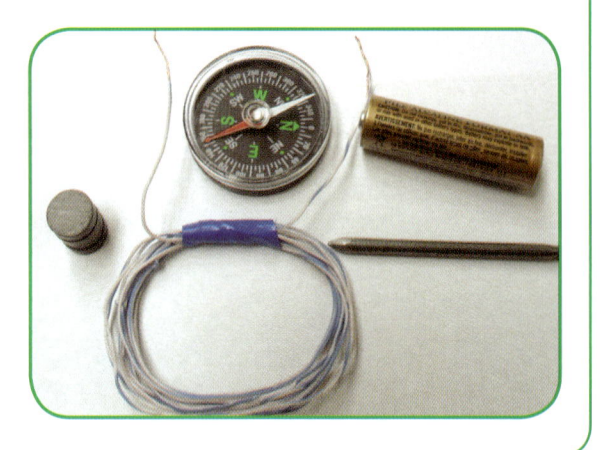

Activity Timeline and Getting Ready

- Activity will take approximately 20 minutes.
- Divide youth into small groups of two or three.
- Provide each team with supplies.
- The insulation on the ends of the wire should be trimmed back so about ½ inch of bare wire is showing. For younger members, you may want to do this before the meeting. Older groups can do this if you provide the tools.

Experiencing

1. Lead the groups in discussion: In the light bulb activity, you used an electrical current to create light. You also felt the heat from the flow of electricity. You know that electricity moves parts in a blender, a drill, or a fan. A common electrical device for movement is a motor. Ask youth to record in their Robotics Notebook and then share with the entire group:
   a. What do you know about how a motor works?
   b. How do you think electricity makes things move?

2. If the concept of magnetic force did not arise from the discussion, share that some believe there is a magnetic force around a wire when the electricity is flowing through the wire. They will be experimenting with wire and batteries to determine if this is true.

3. Each participant should have a piece of the wire.

4. **Test 1** – Using just the plain straight wire, have the groups work together to complete a circuit by touching the ends of the wire to the battery. Then they should see if the wire will magnetically pick up any paper clips or move the compass needle. Check the items by using a magnet to pick up the same items. Have them record what happens in their Robotics Notebook.

5. **Test 2** – Ask the participants to take the two lengths of wire and make two coils by wrapping the wire around a cylinder multiple times.

   a. Another group member should wrap the wire around a nail and tape the coil to the nail.

   b. Another group member should wrap wire around a battery and then slide it off and tape the wrapped wire to hold the coil shape.

   c. Have the group use the two coils to experiment with electricity and magnetism. Have them add this data to the chart they started in Test 1.
   - Can they pick up the paper clips or other items?
   - Can they make the items drop?
   - Can they move the needle in the compass?
   - Have participants record their data and observations in their Robotics Notebook.

6. **Test 3** – Using a toy motor, connect the leads to a battery. Observe how the motor moves (turns). Based on their testing of magnetic force, ask the youth to use their Robotics Notebook to draw or describe how they think the inside of the motor is made or works.

Sharing and Processing

As the facilitator, help guide youth as they question, share, and compare their observations. Before they share with the group, have youth reflect on the activity in their Robotics Notebook. Use more targeted questions as prompts to get to particular points. There is no one right answer.
- Share your observations about the differences between each test – straight circuit and coil.
- Why did making a coil make a difference in the magnet's strength?
- How were you able to let go of the items? If not, why not?
- If you got the compass needle to spin, how did you keep it spinning?
- Have the groups use data from their Robotics Notebook to form a hypothesis about how a motor works. Have each team share its claim with the entire group.

Generalizing and Applying

- Where have you seen or used electromagnets in your house, garage, or school?
- Share other types of electromagnets you've seen. Are they similar or different than the ones you've created?
- What are some things that you could design using electromagnets?
- How does a motor use electricity and electromagnetism?
- Youth can apply what they have learned in Activity G.

Activity G – Can-Can Robot Design Team

Performance Task For Youth

You will plan and engineer a robot with a paper or plastic cup, felt markers, a motor, a battery, wire, an eraser, and tape.

Success Indicator

Youth will be able to design plans for a simple electrical motor-operated robot.

List of Materials Needed

- Robotics Notebook
- Activity Supplies
  – Set of materials for display only. A clear plastic container may be helpful for display.
  – AA battery
  – Small toy motor, about 1.5 to 6 volts
  – Two to four pieces of small wire such as telephone wire or 18-22 gauge insulated copper wire in 6- to 10-inch lengths
  – Several wire strippers if the wire is not stripped
  – Three or four thin felt tip markers of various colors of washable ink
  – One pencil eraser
  – One 8-ounce to 16-ounce paper or plastic cup (Note: Some plastic cups may be harder for the participants to work with, such as in drilling or poking holes.)
  – Poster paper or newsprint

Activity Timeline and Getting Ready

- Activity will take approximately 20 minutes.
- Divide youth into small groups of two or three.

Experiencing

1. Display the materials to be used for this activity. The Design Teams may look at but not touch or play with items in this design stage.
2. Give limited instructions on the design task. Participants are to design a robot that will make drawings on paper using only the supplies provided.
3. Allow Design Teams time for discussion on how to make the robot.
4. Have each Design Team use the Robotics Notebook to plan and draw ideas. The groups should make sketches of their plans.
5. Ask the Design Teams to share the Can-Can Robots they have designed.

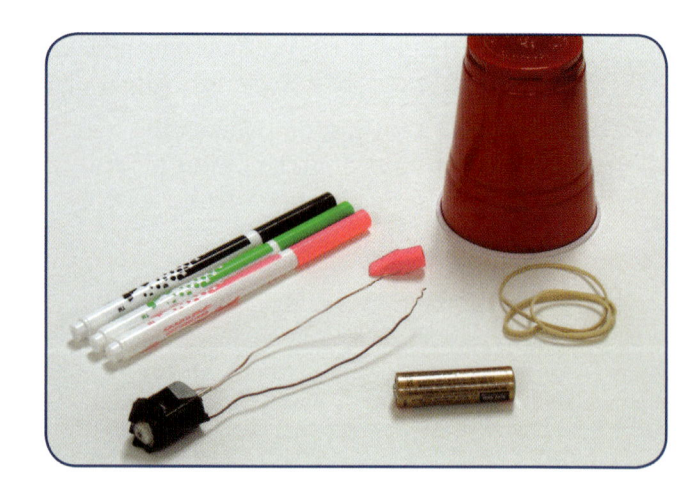

Sharing and Processing

- What type of design does each robot use?
- How can you get the robot to draw?
- How will electricity help the robot draw?
- What are the different ways the robot will draw?

Generalizing and Applying

- What other parts might make it easier to build or use this robot?
- What other things could you design or plan?
- Youth also can apply what they have designed in Activity H.

4-H Junk Drawer Robotics: Robots on the Move • **Module 2: Watt's Up?**

Activity H – Can-Can Robot Build Team

Performance Task For Youth

You will explore how parts can be assembled to build or make more complex things. This activity will encourage you to explore and modify designs using different types of parts to construct a machine.

Success Indicator

Youth will be able to build a simple electrical-motor-operated robot from supplied parts.

List of Materials Needed

- Robotics Notebook
- Activity Supplies

  Sets of materials for each group should include:
  - AA battery
  - Small toy motor, about 1.5 to 6 volts
  - Two to four pieces of inside telephone wire or 18-22 gauge insulated copper wire in 6- to 10-inch lengths
  - Three or four thin felt tip markers of various colors of washable ink
  - One pencil eraser
  - One paper or plastic 8-ounce to 16-ounce cup (Note: Some plastic cups may be harder for the participants to work with, such as in drilling or poking holes.)
  - Poster paper or newsprint
- Toolbox
  - Hand drill with bits
  - Wire strippers
  - Masking tape
  - Rubber bands
  - Safety glasses

Activity Timeline and Getting Ready

- Activity will take approximately 30 minutes.
- Use the same groups from Activity G, Can-Can Robot Design Team.
- Cover the work areas with poster paper or newsprint so the drawing pens won't mark the work surface.
- Solder the lead wires onto the motor contacts and use electrical tape to fasten them to the side of the motor for younger participants. See the 4-H Robotics website at *www.4-H.org/curriculum/robotics* for examples. Older members may be able to do their own soldering if soldering irons are available. Follow the safety note on the use of soldering irons to avoid injuries.

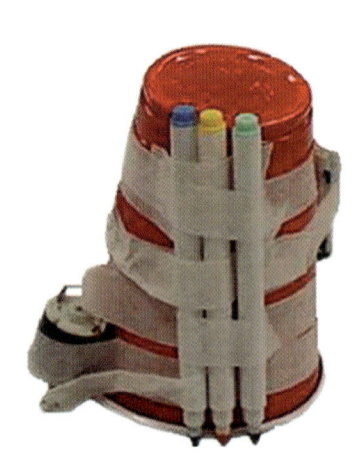

Experiencing

1. Ask the Build Teams to build a robot that will make drawings on paper. (Note: If designs have swinging felt pens, the ink may splatter, so cover nearby objects and clothing.)
2. Use the ideas from the Design Team activity to build the robots.
3. Give each group the parts to be used in building the robots.
4. The Build Teams should do the building, testing, and modifications.
5. When the robots are complete, have the Build Teams present their robots and demonstrate how the robots draw.

Sharing and Processing

As the facilitator, help guide youth as they question, share, and compare their observations. Before they share with the group, have youth reflect on the activity in their Robotics Notebook. Use more targeted questions as prompts to get to particular points. There is no one right answer.
- What designs worked well? (cup up, cup down, cup on side, other)
- What designs did not work as well?
- If the robot made a circle or a spiral drawing, explain why.
- What are some other ways these robots could be made? (e.g., pen stays still and the paper moves, swing arms)
- How could you change (reprogram) the drawing pattern? (Variables, pen spacing, motor speed, location, size of eraser, or other offset vibrator used.)

Generalizing and Applying

- What are some common items that work like our robots? (printers, cell phone vibrate mode, windshield wipers, blender)
- What are some other uses of electromagnetic forces?
- Youth can apply what they have learned in the next module.

Module 3: Get a Move On

Overview of Activities in this Module

To Learn
Activity I – Gear We Go Again
Activity J – Gears and More Gears

To Do
Activity K – Gear Train Design Team

To Make
Activity L – Gear Train Build Team

To Do
Activity M – Es-Car-Go Design Team

To Make
Activity N – Es-Car-Go Build Team

Note to Leader

What is a robot rover? A robot rover is usually either an underwater Remotely Operated Vehicle (ROV) or an Unmanned Ground Vehicle (UGV). These are types of robots that can function away from the operator, can go into places that may be too small or dangerous for a person to fit, or are large and carry heavy loads. Operation of robot rovers may be executed with a long cable, radio control, or other programming.

Rovers are used on the moon, by bomb squads, in mining, for underwater exploration, in dangerous places, and in sweeping kitchen floors. Although robot rovers usually provide the main movement of a robot, other devices such as arms or probes may be needed to complete the task for which it is designed and can be attached to the rover. The energy or power from a motor or other source to drive the rover needs to be controlled and directed. A drive or gear train will help direct the motion, speed, and direction of the movements from the motor. A variety of gears, levers, cams, and other items can be used to form the drive train. We are going to focus on Unmanned Ground Vehicles for this module. In later modules we will explore underwater ROVs.

Toy gears (or any gears) need to be part of a set of gears, meaning they will fit together because they have the same style and size of gear teeth. The sizing of gear teeth is called the teeth pitch. Teeth of the same pitch will fit together.

What you will need for Module 3: Get a Move On

- Robotics Notebook
- Trunk of Junk, see page 8
- Activity Supplies
  - For this module: See the 4-H Robotics website at *www.4-H.org/curriculum/robotics* for sources of wheels, gears, and other items.
  - Sets of gears (from old toys or appliances or buy at science/tech supply stores)
  - Axles (e.g., straws, nails, coat hangers, paper brads, dowels; some plastic coffee stirrer straws fit very well into toy wheels)
  - Toy motors, 1.5 to 12 volt
  - Wheels – various round objects, bottle caps, toy wheels, or disks of different materials, diameters, width, tread and center holes
  - Wires – small gauge 20–28; telephone-type wire can be used
  - Batteries – AA, C, or D, or 6-volt lantern
  - Cardboard, plastic canvas, or other pieces to attach and set up gears
  - Fasteners – paper brads, paper clips, etc.
  - Craft sticks – various sizes with drilled holes
- Things to make or acquire
  - Copies of handout/poster of types of gears on page 52
  - Structure parts (drilled craft sticks), or similar items. See the 4-H Robotics website at *www.4-H.org/curriculum/robotics* for tips on how to make drilled sticks.
  - Use or borrow one or more bicycles that have multi-gear drive systems (5-, 10- 21-speed, etc.)
- Tool box
  - Tape – masking or electrical
  - Low-temperature glue gun
  - Hacksaw blades
  - Hand drill and bits
  - Pliers, punches, and cutters
  - Bench hook or work surface

Timeline for Module 3: Get a Move On

Activity I – Gear We Go Again
- Activity will take approximately 20 minutes.
- Have one or more bicycles for the teams.
- Divide youth into teams of two to four.

Activity J – Gears and More Gears
- Activity will take approximately 40 minutes.
- Get drilled craft sticks, gears, paper brads, straws, mounting backing (cardboard, plastic canvas, etc.).
- Either divide youth into new pairs or use the same teams from Activity I, Gear We Go Again.

Activity K – Gear Train Design Team
- Activity will take approximately 20 minutes.
- Use the same teams from Activity J, Gears and More Gears.

Activity L – Gear Train Build Team
- Activity will take approximately 30 minutes.
- Use the same groups from Activity K, Gear Train Design Team.

Activity M – Es-Car-Go Design Team
- Activity will take approximately 20 minutes.
- Full group for the first part; divide youth into groups of two or four for the second part.

Activity N – Es-Car-Go Build Team
- Activity will take approximately 30 minutes.
- Use the same groups from Activity M, Es-Car-Go Design Team.

Focus for Module 3: Get a Move On

Big Ideas
- Drive trains
- Engineering design
- Robot rovers
- Robots use simple machines.

NSE Standards
- Abilities of technological design
- Form and function
- Motion and forces

STL
- Engineering design
- Core concepts
- Energy and power technologies
- Characteristics of technology
- Problem solving

Performance Tasks For Youth

You will work with a multispeed bicycle to understand gear ratios.

You will assemble and test gear sets to determine the direction of rotation and gear ratios. You also will explore compound gear ratios.

You will design a gear train that will have a gear ratio reduction.

You will build a gear train using compound gears.

You will plan and design a rover with a gear train to make it go really slow and climb a ramp.

You will build a rover that can go as slow as an "Es-Car-Go" (snail) and is able to climb a ramp.

SET Abilities
- Build/Construct
- Collect Data
- Draw/Design
- Hypothesize
- Observe

Life Skills
- Critical Thinking
- Keeping Records
- Sharing

Success Indicators

Youth will understand gear ratios and why a bicycle has different sizes of gears.

Youth will be able to describe a simple drive train using gears and calculate gear ratios, including compound gears.

Youth will be able to design a drive train that can reduce output speed.

Youth will have a compound gear train with low output speed.

Youth will design a plan for using gears in a slow-moving rover vehicle.

Youth will construct a rover that uses a set of gears (gear train) to move slowly with good traction.

Activity 1 – Gear We Go Again

Performance Task For Youth

You will work with a multispeed bicycle to understand gear ratios.

Success Indicators

Youth will understand gear ratios and why a bicycle has different sizes of gears.

List of Materials Needed

- Robotics Notebook
- Activity Supplies
  – Two or three hoops of different diameters – hula hoop, craft wreath ring, ring toss, etc.
  – At least one (1) multispeed bike (as long as the rear wheel has multiple gears). If there are more bikes, allow each group to experiment with one.
  – Tape to mark the wheel
  – Copies of handout/poster of types of gears on page 52

Activity Timeline and Getting Ready

- Activity will take approximately 20 minutes.
- Place the bike upside down (balanced on seat and handle bars) or on a bike stand.
- Move the foot pedals to a vertical or high/low position and then use the tape to mark the top of the rear wheel.
- Measure the diameter of the different gears and crank.

Safety Note:

Many parts on a bicycle — levers, wheels, and other moving parts — can cause injury. When experimenting with the bike gears and wheels, be careful so fingers aren't pinched or caught in the spokes or other moving parts of the bike.

Experiencing

1. Lead discussion with full group about how a robot moves under its own power.
   a. Have youth share what they know about robots.
   b. Ask youth what they know/don't know about robot movement.
   c. Ask participants to give examples of rovers.
   d. Compared to the Can-Can robot, how could gears be used to move robots?

2. Play the Gear Game. Organize youth into two groups with a different number of members in each group. One group will be the "driver gear" and one will be the "driven gear." For small groups of two to four, the members will interlink their elbows (all right or left elbows) to make a star-like shape.
   a. Groups of five to eight might use a hoop or ring to hook their elbows through to allow more members in the circle. And larger groups of 10 or more might use a hula hoop to link their elbows to make an even larger diameter "gear" circle.
   b. The youth will put their non-linked arm straight down along their side, so they each become a "tooth" on the gear. Have the "gears" carefully move together (mesh) so

that one tooth of a gear fits between two teeth of the other gear.

c. Now have the "driver gear" start to turn around its center point, thus having the "driven gear" turn around its center. Have each "gear" count how many times he or she goes around compared to the other "gears." Ask members to make observations such as if the teeth (members) always mesh between the same other teeth. What effect did linking left or right elbows make? If there is time, repeat with different-sized groups and/or hoops.

3. Discussion with the youth: Ask how many of the youth have ridden a bicycle before. What happened when they peddled into the wind or up a hill? What about on a long, flat road? Did they ride a BXM, or single-speed, bike or a 10-speed or multispeed bike? How did that make a difference when riding up that hill?

   a. What are the "speeds" on a bike? Don't they all just have the speed that you peddle the bike? (Note: the term "speed" like "10-speed" or "4-speed" used in some bikes or cars really means the number of different ratios of gears or sprockets allowing various speeds or movement.)

   b. Why do they have so many sprockets (gears) on the back wheel of some bikes? How does changing which gear the chain is on affect your peddling?

4. Have the groups experiment with calculating and changing ratios on a bike. If there are not enough bikes for each group to have its own bike to work with, divide activities to be shared with two or more groups working on a bike. Try not to have more groups than there are gears on the rear wheel (usually from five to seven gears).

5. With the bike upside down, steadily braced, and following safety rules for hands and fingers, have the first group calculate the ratio when the chain is on the smallest gear (sprocket) on the back wheel.

   a. To calculate the ratio, youth should slowly turn the pedals one complete turn and count the amount of turn(s), or part of turn, the rear wheel, makes for the one rotation of the pedals.

   b. (The rear wheel should not be allowed to free wheel or coast beyond peddling movement.) Marking the tire with chalk or tape may help youth in counting the wheel rotations.

6. **Sample ratio calculations:** If the foot pedal rotates one turn and the rear wheel rotates twice, the ratio would be 1 to 2 (1:2) in this example. If the foot pedal rotates one turn and the rear wheel rotates three and a quarter turns, the ratio would be 1 to 3 ¼ (1:3.25). When comparing rotation, the calculation is from the driver to driven.

7. Have the groups calculate the gear ratio of each of the different speeds on the bike. Post results on a poster or chart.

   (**Note:** Some bikes also may have multi-sprockets on the pedal crank. If so, adjust the front cluster to a different gear and then the groups can determine this new ratio for each of the rear gear options.) Once all the gear ratios have been calculated and posted, have the groups compare their findings.

8. **Optional ways to calculate ratios:** If after the youth have used rotations to calculate and recorded all the gear ratios, challenge them to use an alternative way of calculating the same ratios.

   a. Counting the number of teeth on the driven and on the driver gears (sprockets – 15 teeth by 30 teeth).

   b. Measure the diameter of the driven gear and the gear on the crank. (3 inches driven divided by 6-inch driver).

Note: When calculating using the number of teeth or diameter of the gears, you compare the driven to the driver to get the correct ratio. Rear wheel sprocket 15 teeth to pedal crank sprocket 30 teeth would be 15/30 or 1/2 (1:2).

Sharing and Processing

As the facilitator, help guide youth as they question, share, and compare their observations. Before they share with the group, have youth reflect on the activity in their Robotics Notebook. Use more targeted questions as prompts to get to particular points. There is no one right answer.

- What did you notice when you played the gear game? Were each of you evenly spaced to fit together? (pitch of the gears) Did one of the "gears" have to walk (turn) backward? (rotation direction) Did you bump into each other when you meshed and turned? (friction and wear)
- What did you notice when testing the bike ratios? Were there any ratios that were the same or overlapping? What was the difference in using gear sprockets and a chain compared with two gears directly meshing together?
- Why do some bikes, cars, tractors, and other equipment have different-sized gear ratios? How would this affect riding in different conditions? (flat road, up hills, down hills)

Generalizing and Applying

Discuss with the youth where else they would find applications of gears and what they are designed to do or work.

- How have you used your own bike while riding it?
- Describe machines that have or use gears and the function of the size of gears or ratios they use. Ice cream maker? Toy friction motor car? Hand crank egg beater?
- Youth can apply their knowledge in future activities that deal with gears.

Activity J – Gears and More Gears

Performance Task For Youth

You will assemble and test gear sets to determine the direction of rotation and gear ratios. You also will explore compound gear ratios.

Success Indicators

Youth will be able to describe a simple drive train using gears and calculate gear ratios, including compound gears.

List of Materials Needed

- Robotics Notebook
- Activity Supplies
  – Axles (e.g., straws, nails, coat hangers, paper brads, dowels)
  – Structure parts (drilled craft sticks), similar items and/or mounting backing (cardboard, pegboard, plastic canvas, etc.)
  – Gears – various sizes of gears, optional worm and bevel gears from old toys or appliances or buy at science/tech supply stores
  – Copies of handout/poster of types of gears and ratios on page 52
- Toolbox
  – Low-temperature glue gun
  – Small drill bits and hand drill
  – Punches for making holes

Activity Timeline and Getting Ready

- Activity will take approximately 40 minutes.
- Make or obtain drilled craft sticks, gears, backing, and axles.
- Divide youth into teams of two or three.

Experiencing

Step 1

1. Have gears available for participants to set up and test. Ask them to mount two gears on a drilled stick or piece of plastic canvas so that the two gears connect (mesh). Youth should turn one of the gears and observe the other gear. Have members record their observations about how gears function in their Robotics Notebook.

2. Using these two gears, have members count the number of teeth on the driver (the one they turned) gear (input) and the number of teeth on the driven gear (output). They should also record this in their Robotics Notebook and use the chart to calculate the gear ratio for those two gears.

3. Have the teams repeat Step 1 with different sizes of gears.

Step 2

4. Have groups construct a drive train (gear set) that will satisfy the following criteria:
   - Use at least three gears.
   - Driven gear must turn at least three times faster than the first driver gear.
   - Driven gear must turn in the opposite direction of the first driver gear.

5. Ask youth to draw the gear set they designed and try to figure out the gear ratio from the first driver gear to the last driven gear and record the gear set in their Robotics Notebook. What might be a limitation to gear trains that are too fast or slow? (Size of gears has to be too large or too small.)

4-H Junk Drawer Robotics: Robots on the Move • Module 3: Get a Move On

6. Ask a few general questions before moving to the next step:

   a. What do you think gears might be used for? In what tools or machines have you seen gears used? How were they used?

   b. Ask youth how the ratio from the last driven gear to the first driver gear is different from the ratio of just a two-gear set of the same size driven and driver gears? (The ratio is the same; the extra gears in between are just idlers, though direction of rotation might have changed.)

Step 3

7. Ask youth what they know about the meaning of the word "compound." Ask youth what they think is meant by a compound gear. Ask for a volunteer to draw a compound gear on the board or poster. From this sharing, ask how a compound gear could be designed (two gears together or on the same shaft or axle). What might be the advantage of using a set of compound gears? Share that they will now be designing and building a compound gear train.

8. Challenge teams to use craft sticks or other backing to support a gear train that will have at least an 8:1 ratio to either increase or decrease speed. They should use at least one compound gear, but may use more.

9. To get the gears to mesh, they may have to drill new holes in the craft stick for correct spacing and alignment of the gears.

10. Have them draw their gear train design and try to figure the overall ratio for this gear train in their Robotics Notebook. **Note:** The ratio of each two meshing gears is multiplied with the next two meshing gears from the compound gear. (The blue and yellow gears make up a compound gear. To figure overall ratio, calculate the ratio of blue and red gears and then multiply by the ratio of the green and yellow gears.)

11. **Optional:** If available, use a worm gear to increase the ratio, or use bevel gears to change the rotating axle plane.

Sharing and Processing

As the facilitator, help guide youth as they question, share, and compare their observations. Before they share with the group, have youth reflect on the activity in their Robotics Notebook. Use more targeted questions as prompts to get to particular points. There is no one right answer.

- How did the compound gears make a difference in the gear train?
- What was the hardest part of using the compound gears?
- How could you have designed a regular straight set of gears to get the same ratio in your gear train?

Generalizing and Applying

- Where do you think compound gears are used?
- Youth can apply what they have learned in Activity K.

Some common types of gears

Pinion gear – usually a small gear used as a driver. It may be attached directly to a motor.

Spur Gear – the most common type of gear. Different size diameter gears are used together in a set to change speed or rotation direction.

Rack Gear – can be used to convert rotation motion into linear motion, or vise a versa, either with the rack moving or the pinion moving across the rack.

Worm Gear – good for gear reductions as they have one spiral grove or tooth. They also have a locking quality in that the worm can easily turn the driven spur gear but the spur gear cannot make the worm turn. This is a good way to hold robot arms, wheels, or other gears in place or position.

Bevel Gear – useful to change direction of the rotating shafts — usually at 90 degrees — but some are made for other angles.

Compound Gear – two or more gears of different sizes (diameter) connected so they rotate together either on one shaft or as one piece. Used to compound or increase the gear ratios in changing rotational speed or ratio.

Example of how to figure Gear Ratios

Driven (green) spur gear (40 teeth)

Driver (red) pinion gear (12 teeth)

Ratio = Driven gear / Driver gear
Ratio = 40 divided by 10 = 40/10 = 4/1 = 4:1 ratio

Activity K – Gear Train Design Team

Performance Task For Youth

You will design a gear train that will have a gear ratio reduction.

Success Indicator

Youth will be able to design a drive train that can reduce output speed.

List of Materials Needed

- Robotics Notebook
- Activity Supplies
  - For viewing only during design stage: gears, tongue depressors, plastic canvas mesh, brads, straws, skewers, toothpicks, other items for axles
  - Toy motors, 1.5 to 12 volt
  - Structure parts (drilled craft sticks), or similar items
  - Copies of handout/poster of types of gears on page 52

Activity Timeline and Getting Ready

- Activity will take approximately 20 minutes.
- Display material for youth to view during design.
- Divide youth into teams of two to four.

Experiencing

1. Have Design Teams of two to four design a gear train that will:
   a. Be powered by motor and battery (full power).
   b. Use gears to go slower (will be used in Activity M, Es-Car-Go Design Team).
   c. Propel a car (robot) to climb a cardboard ramp at an incline.
2. Display/show materials to be used for this activity. Design Teams may look at but not touch or play with items in this design stage. Designs are restricted only by the supplies you provide or that are available to them.
3. Have each Design Team use the Robotics Notebook to plan and draw their ideas. Groups should make sketches of their plans.

Sharing and Processing

As the facilitator, help guide youth as they question, share, and compare their observations. Before they share with the group, have youth reflect on the activity in their Robotics Notebook. Use more targeted questions as prompts to get to particular points. There is no one right answer.

- Ask each group to share its design.
- Why do you think your gear train will run slowly?
- How do you plan to get the power from the gear train to the mobile robot?

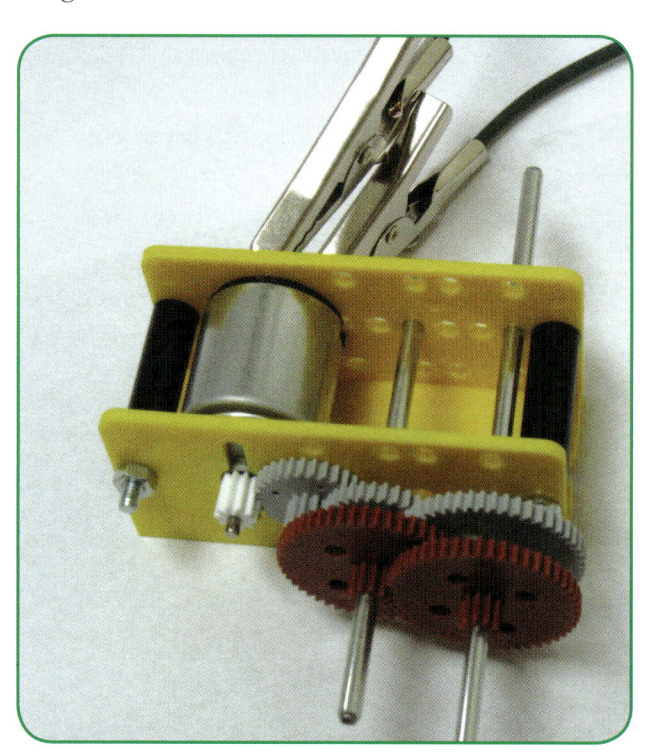

Generalizing and Applying

- What other information would be helpful to complete your design or plan?
- Share where you will try to reduce friction and where you will try to increase friction.

Activity L – Gear Train Build Team

Performance Task For Youth

You will build a gear train using compound gears.

Success Indicator

Youth will have a compound gear train with low output speed.

List of Materials Needed

- Robotics Notebook
- Activity Supplies
  - Junk Drawer material
  - Sets of gears (from old toys or appliances or buy at science/tech supply stores)
  - Axles (e.g., straws, nails, coat hangers, paper brads, dowels)
  - Toy motors, 1.5 volt to 6 volt, and batteries
  - Structure parts (drilled craft sticks), or similar items
  - Copies of handout/poster of types of gears on page 52
- Toolbox
  - Low-temperature glue gun
  - Wires
  - Tape
  - Saw, pliers, scissors
  - Drill bits and hand drill
  - Punches for making holes

Activity Timeline and Getting Ready

- Activity will take approximately 30 minutes.
- Use the same teams from the previous activity.

Experiencing

1. Using the following criteria, build a gear train that:
   a. Is powered by motor and battery (full power).
   b. Uses gears to go slower (will be used in Activity M, Es-Car-Go Design Team).
   c. Can be used to propel a car to climb a cardboard ramp at an incline.
2. Have teams share and demonstrate completed gear trains.

Sharing and Processing

As the facilitator, help guide youth as they question, share, and compare their observations. Before they share with the group, have youth reflect on the activity in their Robotics Notebook. Use more targeted questions as prompts to get to particular points. There is no one right answer.

- Describe what you observed as you built the gear trains for this activity.
- How did the gear trains use different types of parts or amounts of parts?
- What functions do the different parts serve? (Gears? Axles? Spacers? Wheels?)

Generalizing and Applying

- How do you think engineers create gear trains?
- How do you think engineers select the materials to be used in their gear trains?

Activity M – Es-Car-Go Design Team

Performance Task For Youth

You will plan and design a rover with a gear train to make it go really slow and climb a ramp.

Success Indicator

Youth will design a plan for using gears in a slow-moving rover vehicle.

List of Materials Needed

- Robotics Notebook
- Trunk of Junk
- Activity Supplies
  - Sets of gears – use the gear train built in Activity L
  - Axles – straws, nails, coat hangers, paper brads, dowels
  - Toy motors, 1.5 to 6 volt
  - Structure parts – drilled craft sticks or similar items
  - Rubber bands
  - Wheels – various round objects such as bottle caps, toy wheels, or disks of different materials, diameters, widths, treads, and center holes

Activity Timeline and Getting Ready

- Activity will take approximately 20 minutes.
- Divide youth into groups of two to four.
- If desired, use constraint of costs and adapt by adding budget, supplies, and inventory as in Module 1, Activities 1-C and 1-D.

Experiencing

1. Lead discussion with the entire group about using available materials to make a robot that will move under its own power.
   a. Have youth share what they know about robots.
   b. Ask youth about what they know/don't know about robot movement.
   c. Ask participants to give examples of rovers.
2. Have youth form Design Teams of two to four. The teams will design a robot rover that will:
   a. Be powered by a motor and battery (full power).
   b. Use a drive train of gears (may use gear set built in Activity L).
   c. Move slowly (go as slow as youth can make it, like a snail).
   d. Climb a cardboard ramp at an incline.
3. Display materials to be used for this activity. Design Teams may look at but not touch or play with items in this design stage except for the gear train they built in Activity L. Designs are restricted only by the supplies you provide or that are available to them.
4. Have the Design Teams use their Robotics Notebook to plan their ideas. The teams should make sketches of their plans.

Sharing and Processing

As the facilitator, help guide youth as they question, share, and compare their observations. Before they share with the group, have youth reflect on the activity in their Robotics Notebook. Use more targeted questions as prompts to get to particular points. There is no one right answer.

- Ask each group to share its design.
- Why do you think your robot will work?
- How do you plan to get the power to move your robot?
- What other parts might make it easier to use or build this robot?

Generalizing and Applying

- How do you plan to use the gear train you made in Activity L? What types of modifications will you have to make?
- What other information would be helpful to complete your design or plan?
- Share where you will try to reduce friction and where you will try to increase friction.
- Youth can apply what they have designed in Activity N.

Activity N – Es-Car-Go Build Team

Performance Task For Youth

You will build a rover that can go as slow as an "Es-Car-Go" (snail) and is able to climb a ramp.

Success Indicator

Youth will construct a rover that uses a set of gears (gear train) to move slowly with good traction.

List of Materials Needed

- Robotics Notebook
- Trunk of Junk
- Activity Supplies
  - Sets of gears – use the gear train built in Activity L
  - Axles (e.g., straws, nails, coat hangers, paper brads, dowels)
  - Toy motors, 1.5 to 12 volt
  - Structure parts (drilled craft sticks), or similar items
  - Wheels – various round objects, bottle caps, toy wheels, or disks of different materials, diameters, width, tread, and center holes
  - Rubber bands of various sizes, some that fit around the wheels
- Toolbox
  - Low-temperature glue gun
  - Wires and batteries
  - Tape
  - Saw, pliers, scissors
  - Drill bits and hand drill or hole punch

Activity Timeline and Getting Ready

- Activity will take approximately 30 minutes.
- Use the same teams from Activity M, Es-Car-Go Design Team.

Experiencing

1. Have participants build the Es-Car-Go they designed in Activity M, using the following criteria:
   a. The vehicle must be powered by motor and battery (full power).
   b. The vehicle must use a drive train of gears (may use gear train built in Activity L).
   c. The vehicle must move slowly (go as slow as the team can make it, like a snail).
   d. The vehicle must climb a cardboard ramp at an incline.
2. Have teams share and demonstrate the completed vehicles.

Sharing and Processing

As the facilitator, help guide youth as they question, share, and compare their observations. Before they share with the group, have youth reflect on the activity in their Robotics Notebook. Use more targeted questions as prompts to get to particular points. There is no one right answer.
- Describe what you observed as you built the rover for this activity.
- How did the robots use different types of parts or amounts of parts?
- How did the robots differ from each other?
- What functions do the different parts serve? (Gears? Axles? Plates? Wheels?)

Generalizing and Applying

- How do engineers create robots?
- How do engineers determine the design for their robots?
- How do engineers select the materials to be used in their robots?
- Additional challenges:
  – Add a switch to control the robot's movement in turning, stopping, and going forward and backward.
  – Try using an old flashlight to make a rover.
- Youth can apply what they have learned in Module 4.

Module 4: Under the Sea ROV

Overview of Activities in this Module

To Learn
Activity O – Pennies in a Boat
Activity P – Sink or Float

To Do
Activity Q – Sea Hunt Design Team

To Make
Activity R – Sea Hunt Build Team
Activity S – To Make the Best Better Design and Build Team

Note to Leader

So far in the Robots on the Move level of *Junk Drawer Robotics*, we have been working on robot rovers on the ground. With legs (pens) and wheels or tracks, these rovers are part of a group called Unmanned Ground Vehicles (UGV). When robots are designed for use in the air as planes, blimps, or helicopters, they may be referred to as Unmanned Aerial Vehicles (UAV). Robots on the water can be referred to as Autonomous Surface Vessels (ASV). In this module we will explore robots designed for underwater use. These are known as Remotely Operated Vehicles (ROV).

In this module, sinking and floating and how this relates to buoyancy on the surface and underwater will be explored in ROVs. There are different ways to operate remotely. One method is a long cord or tether to provide the control and/or power for the vehicle. This enables the operator to be in a comfortable setting, safe from the elements of the underwater exploration by the ROV.

In other autonomous ROVs, an onboard programmable control with sensors could run autonomously without direct control during the ROV's operation.

Buoyancy, or Archimedes's principle as it is sometimes referred to, means an object has the ability to float. Basically, the force under the object from the fluid (water or air) is at least equal to the force of gravity down on the object. ROVs shouldn't have too much buoyancy because they need to descend into the water. A good ROV will have near to neutral buoyancy, meaning it will neither sink nor rise.

By using a variety of materials and combining items in the ROV, the buoyancy can become near neutral, and the ROV can search underwater with just a small amount of effort. A ROV also can have neutral buoyancy and be just slightly buoyant so if it loses power, its tether could rise to the surface on its own and be repaired. The problem in reaching neutral buoyancy is that it changes as the ROV dives to lower depths when different tools or sensors are added to the ROV frame as more of the tether is lowered into the water.

Most ROVs have thrusters that allow movement in the three axes (X, Y, Z), up and down, forward and reverse, and left and right. Most also have attached lights and cameras and other tools or effectors to gather samples, record data, or perform other work.

Since activities in this module will include having the ROVs dive in a test tank of water (garbage can, water trough, swimming pool, etc.), the design and building stages may take place inside, away from the test tank. This may lead to some original designs not reaching neutral buoyancy very well. Even if the designs seem OK, this is a good time to look at design reiteration. Many times, engineers repeat the process of design, test, and redesign. This reiteration to improve function can be applied to both designs that work and those that don't. An additional activity has been added to allow teams to redesign and then rebuild their Sea Hunt ROV.

What you will need for Module 4: Under the Sea ROV

- Robotics Notebook for each youth
- Trunk of Junk, see page 8
- Activity Supplies
  – Assortment of wood, metal, and plastic items to test for floatability, e.g., paper clips, clothespins, nails, packing peanuts, etc., 10-15 items per group
  – 6-inch squares of heavy-duty aluminum foil, one per group
  – Pennies to use as weights, 100 or more per group (Items from other activities, such as washers, can be used if there are enough.)
  – Metal or plastic coat hangers, one or two per group. Metal ones can be bent and shaped if needed. Hangers for children's clothing will allow more room in the dive tank.
  – 1 ½- to 12-volt small toy motor
  – 6-volt battery
  – 3-4 feet of wire lead (tether), such as telephone cord, one per group
  – Ping-pong balls, 2-4 per group
  – Electrical tape
  – Water dive tank (garbage can, water trough, swimming pool, etc.)
  – Containers to hold water for floating activities, such as 2-liter soda bottles with the tops cut off, one for one or two groups
  – Containers for water such as low, wide cake pans, one for two or more groups
- Toolbox
  – Glue
  – Paper towels to wipe up spills
  – Masking and electrical tape
  – Scissors, one or two per group
  – Pliers, wire cutters, or wire strippers
  – Low-temperature glue gun, two or three to share
  – Hand drill with bits
  – Hacksaw

Timeline for Module 4: Under the Sea ROV

Activity O – Pennies in a Boat
- Activity will take approximately 20 minutes.
- Divide youth into small groups of two to four.
- Provide a piece of heavy-duty aluminum foil and pennies for each group.
- Arrange for a body of water to float the penny boats.

Activity P – Sink or Float
- Activity will take approximately 20 minutes.
- Divide youth into small groups of two to four.
- Provide a selection of items for each team to sort and categorize as "sink," "float," or "don't know."
- Have a body of water to test sinking and floating.

Activity Q – Sea Hunt Design Team
- Activity will take approximately 20 minutes.
- Divide youth into small groups of two or three.

Activity R - Sea Hunt Build Team
- Activity will take approximately 45 minutes.
- Use the same groups from Activity Q, Sea Hunt Design Team.
- Provide specific materials and Trunk of Junk for the teams.
- Have a large, deep tub, tank, or pool for the ROVs to explore.

Activity S – To Make the Best Better Design and Build Team
- Activity will take approximately 20 minutes.
- Use the same groups from Activity R, Sea Hunt Build Team.

**Focus for Module 4:
Under the Sea ROV**

Big Ideas
- Buoyancy
- Engineering design
- Science habits of mind

NSE Standards
- Abilities necessary to do scientific inquiry
- Abilities of technological design
- Motion and forces

STL
- Apply the design process
- Relationships among technologies
- Attributes of design
- Energy and power technologies
- Problem solving

Performance Tasks For Youth

You will learn about weight distribution, surface area, and buoyancy by floating an aluminum foil boat on water and adding pennies as weights.

You will explore the concept of buoyancy, predicting what will float or sink. You will try to float something that normally sinks, and sink something that normally floats.

You will use your knowledge of neutral buoyancy to design an underwater ROV that can be powered to go up and down in a tank of water.

You will build an underwater ROV based on your design.

You will redesign and rebuild your ROV design, making modifications as necessary, based on feedback from testing.

SET Abilities
- Build/Construct
- Communicate/Demonstrate
- Design Solutions
- Draw/Design
- Redesign
- Test
- Use Tools

Life Skills
- Critical Thinking
- Problem Solving
- Sharing
- Teamwork

Success Indicators

Youth will be able to build, test, and redesign an aluminum foil boat that will carry pennies and float on water.

Youth will be able to describe buoyancy in simple terms and combine items to create something that neither floats nor sinks.

Youth will be able to use the concept of neutral buoyancy to design an underwater ROV that can be powered to go up and down in a tank of water.

Youth will build an ROV based on their design.

Youth will use feedback from testing to redesign and rebuild their ROV.

Activity O – Pennies in a Boat

Performance Task For Youth

You will learn about weight distribution, surface area, and buoyancy by floating an aluminum foil boat on water and adding pennies as weights.

Success Indicators

Youth will be able to build, test, and redesign an aluminum foil boat that will carry pennies and float on water.

List of Materials Needed

- Robotics Notebook
- Trunk of Junk
- Activity Supplies
  - Containers for water, such as low, wide cake pans, one for two or more groups to share
  - 6-inch squares of heavy-duty aluminum foil, one per group
  - 100 or more pennies per boat to use as weights (Items from other activities such as washers can be used if there are enough.)

Activity Timeline and Getting Ready

- Activity will take approximately 20 minutes.
- Divide youth into groups of two to four.

Experiencing

1. Design – Have each group use the foil to design a boat to float on the water and hold as many pennies as possible without sinking. Have youth sketch their boat design in their Robotics Notebook. Have each group share its design and explain why group members think it will work.

2. Construction – Have each group build its boat.

3. Testing – One group at a time, start by placing a boat in the water and have youth add pennies until it sinks (destructive testing). Have them count the pennies as they add them to the boat and record in their notebooks and post on poster paper or whiteboard on a chart for all groups.

4. Redesign – Second iteration of constructing and testing – repeat steps 2 and 3.

5. Second iteration of constructing and testing – repeat Step 3.

Sharing and Processing

As the facilitator, help guide youth as they question, share, and compare their observations. Before they share with the group, have youth reflect on the activity in their Robotics Notebook. Use more targeted questions as prompts to get to particular points. There is no one right answer.

- What strategies worked to get more pennies in your boat? Did you put them all on the same spot or spread them out?
- How did you decide to change your design in the redesign phase?
- Why do you think testing and redesigning is important?
- How do you think you could have done nondestructive testing on your boat?

Generalizing and Applying

- What other materials could you use to keep your boat afloat?
- How do you think engineers use the design-build-test-redesign process?
- What are the advantages and challenges of destructive versus nondestructive testing?
- Youth can apply what they have learned in the next activity.

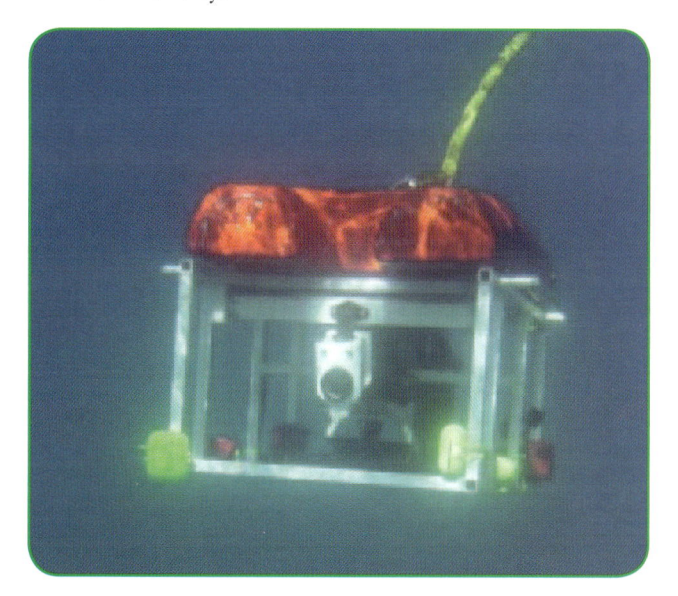

Activity P – Sink or Float

Performance Task For Youth

You will explore the concept of buoyancy, predicting what will float or sink. You will try to float something that normally sinks, and sink something that normally floats.

Success Indicators

Youth will be able to describe buoyancy in simple terms and combine items to create something that neither floats nor sinks.

List of Materials Needed

- Robotics Notebook
- Trunk of Junk
- Activity Supplies
  – Containers such as 2-liter soda bottles with the tops cut off for floating activities, one for one or two groups
  – Wood, metal, and plastic items to test for floatability, such as paper clips, clothespins, nails, packing peanuts, etc., 10-15 per group
- Toolbox
  – General tools as needed
  – Paper towels to wipe up spills

Activity Timeline and Getting Ready

- Activity will take approximately 20 minutes.
- Divide youth into groups of two to four.
- Supply water and pitchers to fill containers for testing.

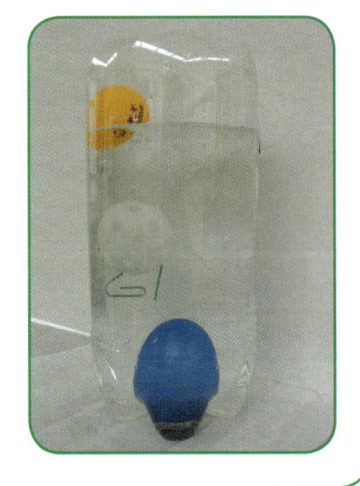

Experiencing

1. Have the groups look at the items that they have to test and categorize them as "sink," "float," or "don't know." They should record their predications in their Robotics Notebooks. Have the groups share their predictions.
2. Ask the groups to test each of the items and record the data in their Robotics Notebook. They should sort the items into piles for sinkers, floaters, and flinkers (items at neutral buoyancy, which means they neither sink or float).
3. Ask the groups to share their results.
4. Now ask the groups to take an item (either one that sank or floated) and to do something to make that item an flinker. Ask them to record in their Robotics Notebook what worked and what didn't work to make that item a flinker.

Sharing and Processing

As the facilitator, help guide youth as they question, share, and compare their observations. Before they share with the group, have youth reflect on the activity in their Robotics Notebook. Use more targeted questions as prompts to get to particular points. There is no one right answer.
- What similarities and differences are there in the items that floated and those that sank?
- Why do you think some items floated and some sank?
- How can some items that normally sink, float?

Generalizing and Applying

- What objects have you seen that float? Why do you think they float?
- Youth can apply what they have learned in the next activity.

CAREER CONNECTIONS

Career Connection 4: Iterative Design Process

Engineers use scientific knowledge to develop safe and economical solutions to real-world problems. When building these products, they follow a basic process of design and creation that is repeated and refined as they work to solve the problem. This repeating and refining is known as an iterative process. The first idea, design, or item is just one step to a final, more complete and developed solution, one step that is repeated over and over until the best results have been obtained.

When presented with a problem or an opportunity, an engineer will first ask questions. These questions help an engineer research a problem, constraints, and objectives. For example, engineers may ask, "What exactly is the problem? What have others done?" Such inquiries allow an engineer to define and state a problem.

Once engineers have identified a problem and researched its background, they begin to imagine the possibilities. They generate ideas and possible solutions. This creative process is essential to develop solutions.

After narrowing down their ideas, engineers evaluate and compare possible solutions so they can plan their next steps. Engineers often use notebooks to write lists of needed resources and draw diagrams of designs. They might also look for other people to help them. Cooperation is an important part of the engineering process. Next, engineers build their device. Using their plan, they use materials to physically craft their idea.

After engineers have a working model, they test and redesign the technology. Engineers know that there is always room for improvement. They communicate and discuss the results of their tests. They modify their device and retest it. This step in the process ensures that engineers create the best possible product.

Finally, engineers communicate the final product and share their solutions with others.
- How have you used an iterative process to solve a problem? Describe what you did.
- What are some products that you think need more iterations to make them better?
- How does iteration help an engineer design better products?

Adapted from the Engineering is Elementary project and Graphic Define on-line magazine (Issue 1 Iterative Design).

Activity Q – Sea Hunt Design Team

Performance Task For Youth

You will use your knowledge of neutral buoyancy to design an underwater ROV that can be powered to go up and down in a tank of water.

Success Indicator

Youth will be able to use the concept of neutral buoyancy to design an underwater ROV that can be powered to go up and down in a tank of water.

List of Materials Needed

- Robotics Notebook
- Trunk of Junk
- Activity Supplies

    For display only, special items for this design:
    – Heavy duty 6-inch squares of aluminum foil, one per group
    – Pennies to use as weights, 100 or more per group (Items such as washers from other activities can be used if there are enough.)
    – Metal or plastic coat hangers, one or two per group. (Metal ones can be bent and shaped if needed. Hangers for children's clothing may allow more room in the dive tank.)
    – 1.5- to 12-volt small toy motor
    – 6-volt battery
    – 3-4 feet of wire lead (tether), such as telephone cord, one per group
    – Ping-pong balls, two to four per group
    – Electrical tape

Activity Timeline and Getting Ready

- Activity will take approximately 20 minutes.
- Divide youth into groups of two or three.

Experiencing

1. Set out the specific materials for use in the ROVs for display only. Design Teams may look at — but not touch or play with — items in this design stage.

2. Have teams design an ROV. Each ROV must fulfill the following:

    a. Be designed to have neutral buoyancy.

    b. Have a tether for motor control.

    c. Be able to move in one of the three coordinate (X, Y, Z) directions, up and down.

3. Have teams use their Robotics Notebook to develop and plan the design for the ROV. After designing it, have the groups share their ROV designs.

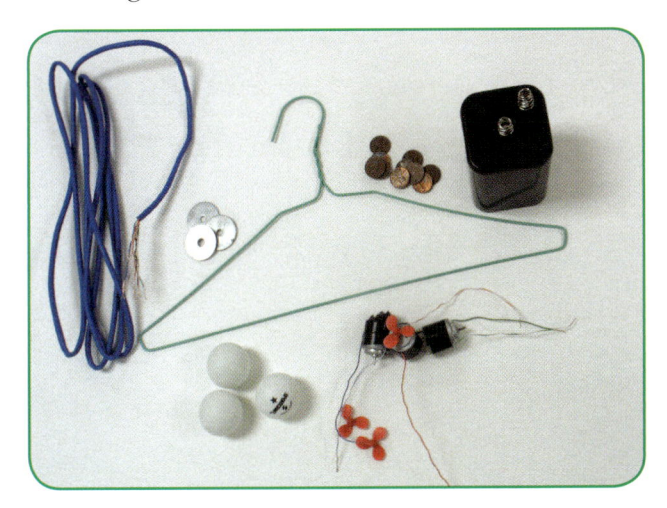

Sharing and Processing

As the facilitator, help guide youth as they question, share, and compare their observations. Before they share with the group, have youth reflect on the activity in their Robotics Notebook. Use more targeted questions as prompts to get to particular points. There is no one right answer.
- What was helpful in making your design?
- What do you think are the strengths of your group's design?
- Why do you think it will be a flinker?

Generalizing and Applying

- How is your design like a ship? Like a submarine?
- Youth can apply what they have learned in the next activity.

Activity R – Sea Hunt Build Team

Performance Task For Youth
You will build an underwater ROV based on your design.

Success Indicator
Youth will be able to build a ROV based on their design.

List of Materials Needed
- Robotics Notebook
- Trunk of Junk
- Activity Supplies
  - Metal or plastic coat hangers, one or two per group. (Metal ones can be bent and shaped if needed. Hangers for children's clothing may allow more room in the dive tank.)
  - 1.5- to 12-volt small toy motor
  - 6-volt battery
  - 3-4 feet of wire lead (tether), such as telephone cord, one per group
  - Toy propeller, 1 to 2 inches in diameter
  - Ping-pong balls, two to four per group
- Toolbox
  - Electrical tape
  - Drill and bits
  - Hacksaw
  - Glue gun and glue sticks

Activity Timeline and Getting Ready
- Activity will take approximately 45 minutes.
- Use the same groups from Activity Q, Sea Hunt Design Team.
- Secure a body of water to run your ROVs. A swimming pool, water trough, plastic container, kiddie pool, trash can, or other containers can be used to float your ROVs.

Experiencing
1. Set out the materials. Have the teams build the ROVs that they designed. Each ROV must fulfill the following:
   a. Be designed to have neutral buoyancy.
   b. Have a tether for motor control.
   c. Be able to move in one of the three coordinate directions, up and down.
2. The members should record actions and modifications in their Robotics Notebook.
3. After building the ROV, have the groups share the ROVs they built.
4. Test the ROVs in a water tank or pool.
5. Have the youth note modifications or changes that need to be made on the design and building of the ROV in their Robotics Notebook.

Sharing and Processing

As the facilitator, help guide youth as they question, share, and compare their observations. Before they share with the group, have youth reflect on the activity in their Robotics Notebook. Use more targeted questions as prompts to get to particular points. There is no one right answer.

- What surprised you about how your ROV acted?
- What worked well with your ROV?
- What did not work so well?
- What tests did you try?

Generalizing and Applying

- Based on your tests, what changes do you plan to make to improve the ROV?
- What would help improve your design?
- What materials would help improve your design?
- Youth can apply what they have learned in the next activity.

Activity S – To Make the Best Better Design and Build Team

Performance Task For Youth

You will redesign and rebuild your ROV design, making modifications as necessary based on feedback from testing.

Success Indicator

Youth will be able to use feedback from testing to redesign and rebuild build their ROV.

List of Materials Needed

- Robotics Notebook
- Trunk of Junk
- Activity Supplies
  - Metal or plastic coat hangers, one or two per group (Metal ones can be bent and shaped if needed. Hangers for children's clothing may allow more room in the dive tank.)
  - 1.5- to 12-volt small toy motor
  - 3-4 feet of wire lead (tether), such as telephone cord, one per group
  - Toy propeller, 1 to 2 inches in diameter
  - Ping-pong balls, two to four per group
  - 6-volt battery
  - Water tank
- Toolbox
  - Electrical tape
  - Drill and bits
  - Hacksaw

Activity Timeline and Getting Ready

- Activity will take approximately 20 minutes.
- Use the same groups from Activity Q, Sea Hunt Design Team.

Experiencing

1. Have the teams share what they learned from testing their ROVs. What has to be done to improve their function?
2. Based on the test results, have them redesign and rebuild their ROV. The ROVs should:
   a. Be designed to have neutral buoyancy.
   b. Have a tether for motor control.
   c. Be able to move in one of the three coordinate (X, Y, Z) directions.
3. After redesigning and rebuilding the ROVs, the groups should retest them in the water tank.
4. The youth should record actions and modifications in their Robotics Notebook.

Sharing and Processing

As the facilitator, help guide youth as they question, share, and compare their observations. Before they share with the group, have youth reflect on the activity in their Robotics Notebook. Use more targeted questions as prompts to get to particular points. There is no one right answer.
- What worked well with your revised ROV?
- What helped improve your design?

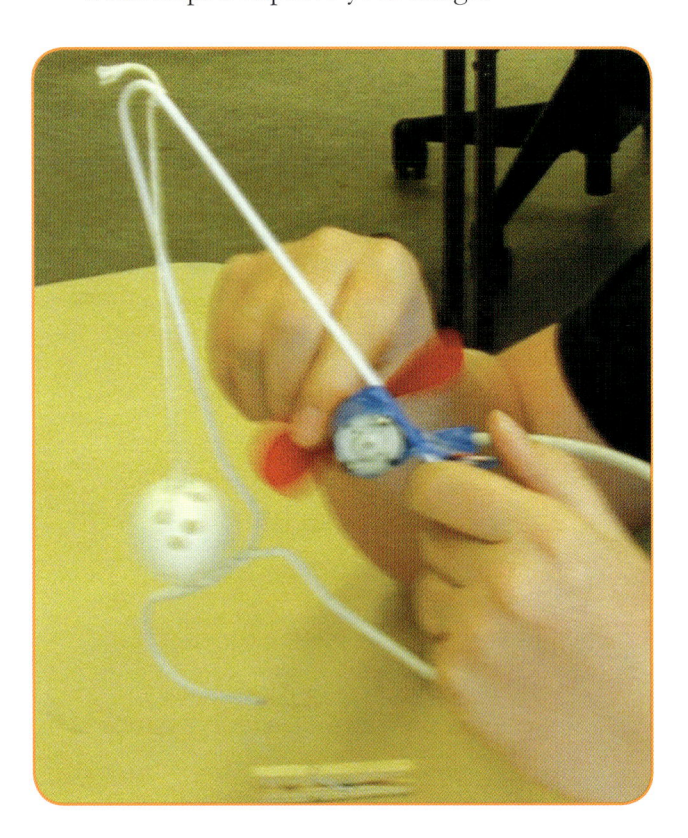

Generalizing and Applying

- What other materials could you use to make an ROV?
- What could your ROV be used for?
- Additional Challenges:
   1. Try adding two more motors so your ROV can move in all three axes.
   2. Try adding other tools to your ROV, such as a flashlight or magnet, and perform a task.
- Youth can apply what they have learned in the next level of this robotics curriculum.

What's Next?

You have completed learning about robot rovers and systems. You learned about friction, gear sets, electricity, and buoyancy. You applied what you learned as you designed a walking can-can robot, an electric-powered es-car-go rover, and an underwater ROV. You used the tools of technology to build and assemble the robots you designed. You were a scientist as you learned, an engineer as you designed, and a technician as you built these robots.

In *Junk Drawer Robotics Level 3, Mechatronics,* you can continue to learn, to do, and to explore what makes robots tick! You will explore how robots use sensors to collect data about their environment. You will investigate how electronic controls help direct a robot. You will learn how to set up a program to code information and create instructions for your robot to follow. These are all part of the next level in *Junk Drawer Robotics*.

If you missed *Junk Drawer Robotics Level 1, Give Robots a Hand,* you may want to try learning about, designing, and building robotic arms that are powered by air.

You also can check out the other 4-H Robotics tracks on Virtual Robotics and Platform Robotics. In these tracks, you can design and build robots on-line or use commercial robotics kits to build and program robots using a computer.

Learn more by visiting the website: *www.4-H.org/curriculum/robotics*.

Glossary

Module 1 – Get Things Rolling

- **Friction** – surface resistance to motion; a force that decreases velocity between two moving objects (Sliding, rolling, and fluid friction have different impacts on objects.)
- **Kinetic** – motion energy that an object possesses
- **Mass** – the quantity of matter (The mass remains constant, but the weight of an object will vary depending on gravity.)
- **Movement** – an object that changes position, measured with velocity
- **Thermal** – heat energy an object possesses
- **Weight** – the force that gravity exerts upon an object
- **Wheel** – a circular disk that revolves on an axis

Module 2 – Watt's Up?

- **Battery** – a device that can store electrical energy that can be used at a later time
- **Circuit** – the complete path of an electric current, typically consisting of a power source (battery), a load (bulb), and an on/off switch
- **Electricity** – a fundamental force consisting of positive and negative charges that provide electric current and power
- **Motor** – a machine that converts electrical energy into mechanical energy
- **Switch** – a device for opening or closing a circuit
- **Torque** – how hard something is rotated or twisted

Module 3 – Get a Move On

- **Axle** – a central shaft for a rotating wheel or gear
- **Cam** – a rotating piece in a drive train used to transfer rotary motion into linear motion
- **Compound Gear** – two gears fixed together so they rotate on the same shaft at the same speed
- **Drive Train** – a group of gears connected to control speed, direction, and power from a motor to outputs such as axles, wheels, pulleys, and cams
- **Energy** – capacity of a system to perform work (Energy exists in several forms including kinetic (movement), thermal (heat), sound, light, and others.)
- **Gear** – a component within a drive train/transmission that uses teeth to transmit mechanical force to another gear or device
- **Lever** – a rigid object used with a pivot to increase the mechanical force applied to an object
- **Pitch** – the distance between points on two teeth on a gear
- **ROV** – remotely operated vehicle; a type of robot that can function away from the operator
- **UGV** – unmanned ground vehicle; a surface ground vehicle that can function away from the operator and can go into places that may be too small or dangerous for a person to fit; also can be large and carry heavy loads